Smooth Topological Design of Continuum Structures

Smooth Topological Design of Continuum Structures focuses on the use of a newly-proposed topology algorithm for structural optimization called Smooth-Edged Material Distribution for Optimizing Topology (SEMDOT).

The book presents the basic theory of SEMDOT and explains connections between this method and the corresponding optimizers. The method is used to address the long-standing jagged edge problem facing classical topology optimization algorithms and is presented as a perfect tool for combining additive manufacturing and topology optimization, which are increasingly coupled together to produce new part designs. A range of representative case studies are also included to illustrate applications.

This serves as a textbook and reference for graduate and senior undergraduate students in the area, as well as engineers in the structural optimization field. Full modifiable MATLAB codes are available online.

Smooth Topological Design of Continuum Structures

Yun-Fei Fu and Bernard Rolfe

CRC Press is an imprint of the
Taylor & Francis Group, an **informa** business

First edition published 2025
by CRC Press
2385 NW Executive Center Drive, Suite 320, Boca Raton FL 33431

and by CRC Press
4 Park Square, Milton Park, Abingdon, Oxon, OX14 4RN

CRC Press is an imprint of Taylor & Francis Group, LLC

ISBN: 978-1-032-61249-2 (hbk)
ISBN: 978-1-032-63443-2 (pbk)
ISBN: 978-1-032-63444-9 (ebk)

DOI: 10.1201/9781032634449

Typeset in CMR10 font
by KnowledgeWorks Global Ltd.

Access the Support Material: www.routledge.com/9781032612492

To the women of our lives:
Jie Gong
and
Siria Natera

Contents

Foreword

Topology optimization has emerged as a crucial and vibrant area of research, demonstrating substantial scientific impact and practical application across a wide spectrum of multidisciplinary fields ranging from aerospace, automotive, mechanical, structural, manufacturing, materials to biomedical engineering. As a computational design approach, topology optimization enables to optimize material distribution within a given design domain, leading to lighter, stronger, and more efficient structures. Although a number of effective topology optimization algorithms have been established and some even included in commercial codes or open-source programs to date, recent technological advances, represented by additive manufacturing, provide new research opportunities yet present new challenges on topology optimization, from algorithmic development to innovative applications.

This monograph aims to introduce a Smooth-Edged Material Distribution for Optimizing Topology (for short SEMDOT) technique, which represents an innovative approach within the realm of topology optimization, specifically targeting on the issues of jagged edges that commonly arise in various density-based methods. These jagged material interfaces can pose considerable difficulties when transitioning from a computational design model to actual manufacturing processes, especially in additive manufacturing where smooth transfer from material layout to real part is crucial for maintaining design optimality and structural performance. Unlike conventional post-processing techniques that focus on modifying the topological output after been generated, SEMDOT takes a proactive strategy by integrating smooth edge layouts directly into the optimization process, inherently exhibiting smoother transitions from design to fabrication and better ensuring optimal performance and manufacturability.

The authors, Prof. Bernard Rolfe and Dr. Yun-Fei Fu, have been working on topology optimization over the recent years. This book represents their latest research on a new topology optimization technique, which particularly facilitates additive manufacturing in various applications. Despite its relatively smaller population compared to other countries, Australia has made significant contributions on the development and advancement of topology optimization since Michelle era in early 1900s. This work demonstrates Australia's continuous contributions to the field.

The book is considered a good reference for college students, researchers, and engineering practicians who are interested in topology optimization. It provides sufficient details of mathematical fundamentals, numerical algorithms, and computer programs as well as many illustrative examples for teaching, research, and development.

I would encourage our community to consider this book and its approaches.

Professor Qing Li
The University of Sydney, Australia
Director, Center for Advanced Materials Technology (CAMT)
President, International Society of Structural and Multidisciplinary Optimization (ISSMO)

Preface

Welcome to the world of SEMDOT. Thank you for reading our text.

Aim of the Book

This book is a culmination of our work developing a topology optimization algorithm to handle removing or reducing support structures in additive manufacturing, which was extended to include smooth surfaces to become Smooth-Edged Material Distribution for Optimizing Topology (SEMDOT). We hope this book will provide support for your teaching or research of topology optimization to allow you to explore optimization of new applications or situations where your surfaces are complex and smooth. Our primary aim was to make a useful textbook that can be used.

In addition, our method addresses the long-standing jagged edge problem with the classical topology optimization algorithms. In this book, we will present the basic theory of SEMDOT as well as explaining how to link the method with the corresponding optimizers. Importantly, we will present some representative case studies that can be followed by the readers.

Another motivation has been the emergence of topology optimization as a heavily used tool by design engineers to reduce the weight of parts and assemblies. Moreover, additive manufacturing has become so ubiquitous, it enables the production of parts with complex geometry. Additive manufacturing and topology optimization are increasing being coupled together to produce new part designs, and there is a need for a tool to combine these two subjects. SEMDOT is a perfect tool for that combination.

How to Use This Book

We envisaged that this book can be used for a senior undergraduate course on topology optimization or a graduate course as part of a solid mechanics or a structures program. This book can also be used as a reference book for PhD and master's students and as a training or learning text for engineers in the structural optimization field. As this book provides different topology optimization case studies, readers can focus on the chapters that are closely related to their real demands. Each chapter will provide the full MATLAB code for the readers to reproduce the results, and the provided codes can be modified by academics for their potential deeper research problems.

The text can provide 35 hours of lecturing time, with an offering of 3 class and tutorial hours per week over a 15-week term for a total of 45 hours total course time. The authors' suggestion is to spend the early weeks building the basics of topology optimization. Weeks 1 to 3 would cover the introduction to topology optimization and the basics of the classical topology optimization algorithms (Chapters 1 and 2). Weeks 4 and 5 would provide the fundamentals of the SEMDOT (Chapter 3). Week 6 could discuss the use of optimizers and present some standard stiffness maximization problems (Chapters 4 and 5). A theory assessment piece could be provided to the students on the first five chapters work. The remainder of the course is then centered on various applications. The authors suggest spending a week on each application topic of interest with a 1- or 2-hour lecture and a 2-hour tutorial based on the provided MATLAB code for each chapter.

Acknowledgments

Australia has a strong tradition of innovation in topology optimization, and it is always important to recognize the shoulders of the "giants" upon which we stand. Australian topology optimization research has been led by Emeritus Prof. Grant Steven (The University of Sydney), Prof. Mike Yi-Min Xie (RMIT University), Prof. Qing Li (The University of Sydney), and Prof. Xiaodong Huang (Swinburne University of Technology). We are happy to build upon this tradition with our new method.

We are deeply indebted to Dr. Kazem Ghabraie (Deakin University, Australia), Prof. Xiaodong Huang (Swinburne University of Technology, Australia), and Dr. Kai Long (North China Electric Power University, China). They significantly contributed to the development of the new topology optimization algorithm introduced in this book. Without the help from Dr. Kai Long, this book cannot be finished on time.

We are grateful to Dr. Zongliang Du (Dalian University of Technology, China), Dr. Hao Li (Kyoto University, Japan), and Dr. Zhi Li (RMIT University, Australia). They provided valuable guidance and some results for Chapter 2.

We would like to thank Prof. Matthijs Langelaar (Delft University of Technology, The Netherlands) for the invaluable guidance on his additive manufacturing filter, which is used in Chapter 10.

We wish to thank Dr. Wenke Qiu (Huazhong University of Science and Technology, China) for the suspension support rod case in Chapter 5, Dr. Tiago Ribeiro (University of Beira Interior, Portugal) and Dr. Luís Bernardo (University of Beira Interior, Portugal) for the steel beam splice connection case in Chapter 5, Dr. Shuzhi Xu (Osaka University, Japan) for the improvement of the additive manufacturing filter in Chapter 10, Dr. Jiye Zhou (Deakin University, Australia) for boundary validation results in Chapter 9 and Appendix B, Dr. Osezua Ibhadode (University of Alberta, Canada) for the industrial frame case in Chapter 10 and Appendix C, and Dr. Minyan Liu (Northwestern Polytechnical University, China) for the discussions on initial designs in Chapter 11.

We thank Dr. Louis N. S. Chiu (Monash University, Australia) for his support with the prototype printing in Chapters 5 and 10 and the fabrication failure graph in Chapter 10, Dr. Jingchao Jiang (University of Exeter, UK) for the discussions on additive manufacturing technologies for Chapter 10, and Dr. Ali Zolfagharian (Deakin University, Australia) for the compressive testing of the lattice structure in Chapter 11.

We appreciate Dr. Joe Alexandersen (University of Southern Denmark, Denmark), Dr. Jie Gao (Huazhong University of Science and Technology, China), Prof. Jikai Liu (Shandong University, China), Dr. Yuan Liang (Dalian University of Technology, China), and Dr. Daicong Da (Boise State University, United States) for the discussions on different topology optimization problems, which are very valuable for this book.

We would like to thank Prof. Krister Svanberg (Royal Institute of Technology, Sweden) for providing the MMA optimizer codes.

Dr. Yun-Fei Fu would like to thank his postdoctoral supervisors Prof. Ahmed Qureshi and Prof. Hani Henein at University of Alberta (Canada). Also, Dr. Yun-Fei Fu appreciates his postdoctoral supervisor Dr. Johannes Reiner at Deakin University (Australia).

Last but by no means least, we thank Tony Moore and Aimee Wragg from the CRC press for their untiring efforts in the production, checking, and editing of the book.

Authors

Dr. Yun-Fei Fu is a researcher in structural optimization. His research interests are lightweight design, topology optimization, additive manufacturing, design for additive manufacturing, and simulation of fiber reinforced composites. His qualifications include a bachelor's degree from the Liaoning Technical University (China) in 2012, master's degree from the Liaoning Technical University (China) in 2014, and a PhD from the Deakin University (Australia) in 2021. He is currently a Postdoctoral Fellow at University of Alberta, Canada. He has published over 20 refereed research articles.

Prof. Bernard Rolfe is a long-time researcher in the automotive sector where he investigates novel manufacturing processes combined with new and advanced materials. He is a Professor of Advanced Manufacturing at the School of Engineering, Deakin University. He has a combined Economics and Engineering degree with honors in 1995 from the Australian National University (ANU), and a PhD in 2002 on Advanced Manufacturing (ANU). He started as a Lecturer at Deakin in 2005, he supervised Deakin's winning submission for the global centennial design competition to devise a Ford Model-T for the 21st century in 2008. His research team found 13% weight savings for pick-up truck canopy in 2014. His research group has spent the past two decades examining the forming of advanced sheet metals primarily for the automotive sector, and over the past decade he has been investigating design and topology optimization for additive manufacturing. His current research focus is the design and forming of lightweight structures. He is also exploring design techniques for the production and repair of parts using additive manufacturing.

List of Figures

List of Tables

1

Introduction

This chapter introduces the basic concepts of optimization, structural optimization, and structural topology optimization. The explanations are given step by step to take the reader through the fundamentals. The organization of the book's chapters are provided for ease of reading.

1.1 Prologue

Optimization is finding the best choice on the basis of a performance criterion or function from a set of available alternatives. A typical optimization will start from an initial guess and iteratively progress toward a better performance choice by continually choosing to improve the situation given some external constraints. In simple terms, the slope or rate of change of the performance function will indicate which way to move to a better situation. The movement in the search space is governed by the type of parameters that make up the search space. If some of the parameters are discrete, that is, have gaps between values in the parameter's set, then the optimization is called discrete optimization. Otherwise, the parameters are all continuous; that is, there are no gaps between values, and this is called continuous optimization. The types of search methods to find the best solution will vary based on the type of optimization (discrete or continuous). In optimization, mathematical principles and methods are used to solve quantitative problems in a wide range of fields, including engineering, physics, biology, and economics.

Structural optimization is an important branch of optimization. It is one of the most intensively explored research areas in engineering, ranging from theoretical research to applications [122]. According to Rozvany [152], in 1870 Maxwell obtained the first optimal structure [129], and this was extended by an Australian mechanical engineer, Michell, in 1904 to derive optimality criteria for the layout of light weight trusses [130]. In the 1970s, Rozvany and his group extended Michell's theory to grid-like structures, for example, trusses, grillages, and shell-grids [152]. Since the beginning of the 21st century, with the rise of computational modelling, structural optimization has qualified as an important simulation-driven design technique that is capable of identifying and exploring the optimal design of load-carrying structures with reasonable structural features. Bendsøe and Sigmund [17] suggest structural optimization can be separated into three basic approaches: topology optimization (Figure 1.1a), shape optimization (Figure 1.1b), and size optimization (Figure 1.1c). In Figure 1.1, the initial optimization problems are shown at the left-hand side, and the optimized structures are shown at the right-hand side. Topology optimization is a structural optimization method that searches for the best material arrangement within a user-defined space for a given set of loads, conditions, and constraints. Shape optimization generally modifies the boundaries of the structure to improve the structural performance while satisfying given constraints. Size optimization finds the best design feature or parameter dimensions of the components whose shapes have already been determined.

DOI: 10.1201/9781032634449-1

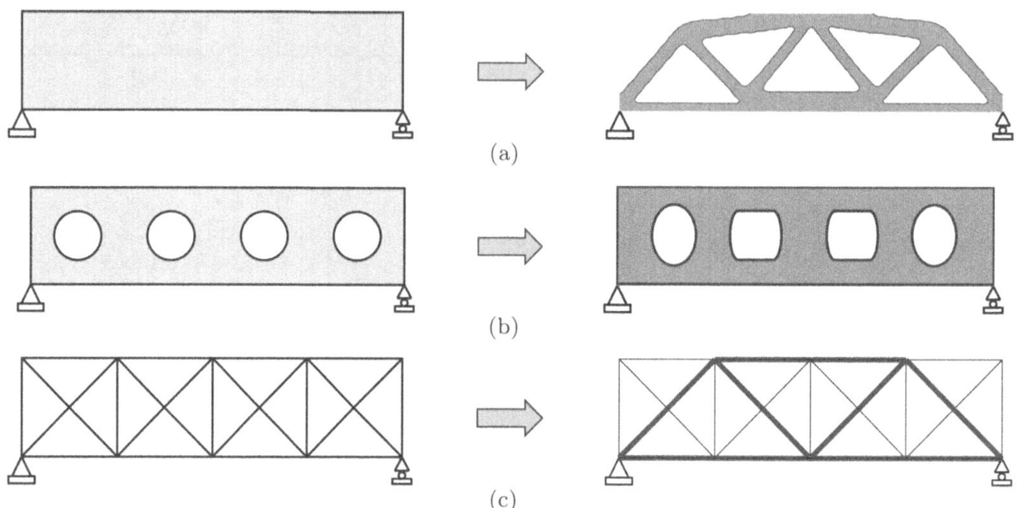

FIGURE 1.1
Three basic approaches of structural optimization: (a) topology optimization, (b) shape optimization, and (c) size optimization.

1.2 Structural Topology Optimization

There are two basic types of topology optimization, discrete and continuum, on the basis of the type of parameters that make up the search space [40]. This book focuses on topology optimization of continuum structures. Structural topology optimization was started from the work by Cheng and Olhoff [30] and has been broadly investigated since the seminal paper by Bendsøe and Kikuchi [19]. Topology optimization essentially separates the design space into solids and voids where the voids can be created, merged, and split during the optimization, and therefore a much larger design space can be explored and improved structural performance can be expected in comparison with the shape and size optimization approaches. Topology optimization has been widely used for mechanical design and weight reduction in various industries (for example, aerospace, automotive, architecture, and medical implants) and academia [163].

Since the 1990s, there have been several classic topology optimization algorithms proposed, including the Solid Isotropic Material with Penalization (SIMP) method [158], the Bi-directional Evolutionary Structural Optimization (BESO) method [81], the level-set method [183], and the Moving Morphable Component (MMC) based method [67]. SIMP and BESO are the most popular element-based approaches. Both methods are able to create holes freely or eliminate unnecessary materials across the design domain, and their final solutions are less dependent on the initial guess configuration. The level-set method is one of the representative boundary-based methods that capture the topological change by merging holes. The MMC-based method is one of the representative feature-driven methods that implements topology optimization by optimizing the arrangement of morphable structural components. These algorithms are widely used in solving different optimization problems, including stiffness, fluid and heat transfer, dynamic, stress constrained, thermoelastic, multi-material, failure, buckling, and large-scale problems [178].

Because of high-design freedom, topologically optimized structures are generally characterized with a sophisticated geometric configuration, which makes the fabrication of these structures very difficult by means of some conventional manufacturing approaches such as machining, casting, or the combination of them. Therefore, most of structures obtained by topology optimization were regarded as conceptual designs in the past. Additive Manufacturing (AM) or 3D printing, a layer-by-layer material deposition or solidification process, is capable of reducing the geometric complexity restriction, and hence topological designs can be easily fabricated in AM [117]. There is no doubt that AM contributed to the rapid development of topology optimization since AM can exploit the full potential of topology optimization. Even though AM is a very powerful technology in fabricating sophisticated geometries, it still has some unique limitations, which should be addressed when carrying out the topological design. Therefore, the limitations such as the minimum overhang angle are considered in topology optimization to improve the manufacturability of optimized structures [221]. In addition, topology optimization for 4D printing is being developed rapidly in recent years [49, 209]. 4D printing can fabricate an object that can change its color, volume, and shape under different environmental stimuli such as heat, humidity, and light, which can provide more design freedom for topology optimization [172].

For all topology optimization algorithms, they have both advantages and disadvantages in solving different structural optimization problems. There is no algorithm that can solve all the optimization problems perfectly, especially when more and more new challenging optimization problems arise. Therefore, new algorithms are still needed in the topology optimization community, with the aim of solving some of the most challenging optimization problems in a simple yet effective way.

1.3 Book Organization

For ease of reading, the remainder of the book is organized as follows:

Chapter 2 briefly introduces some representative topology optimization algorithms, including SIMP, BESO, level-set, and MMC-based methods.

Chapter 3 demonstrates the algorithm mechanism of a newly developed topology optimization tool named Smooth-Edged Material Distribution for Optimizing Topology (SEMDOT).

Chapter 4 discusses two optimizers that are used to update the topological configuration iteratively.

Chapter 5 presents stiffness maximization problems.

Chapter 6 shows compliant mechanism design problems.

Chapter 7 solves both steady and transient heat conduction problems.

Chapter 8 discusses two dynamic cases, which are natural frequency maximization and dynamic compliance minimization problems.

Chapter 9 demonstrates how to conduct the maximum stress minimization problems.

Chapter 10 shows some 3D cases regarding self-supporting design for additive manufacturing.

Chapter 11 presents two representative homogenization topology optimization problems.

Finally, Chapter 12 comprehensively summarizes the book.

This is a note to the readers that the authors have made all examples consistent using the standard SI (mm) units [mm, N, tonnes (10^3 kg), s, MPa, tonne/mm^3]. Where another unit set is used, this will be explicitly mentioned in the text. The use of standard units implies that the authors will not provide units in the text because the units are implied from the type of variable used consistent with the SI (mm) units set. In terms of theoretical research, the readers are also recommended to use a scaling factor only for the objective function values to reduce or increase the final numbers or change to other desired units. This will not affect the final topological configuration. However, for engineering applications, the readers should strictly follow the consistent units.

2

Topology Optimization Algorithms

This chapter introduces the reader to the four classic representative topology optimization algorithms: Solid Isotropic Material with Penalization (SIMP), Bi-directional Evolutionary Structural Optimization (BESO), Level-Set Methods (LSMs), and Moving Morphable Component (MMC) based methods. The reader will be taken through how each algorithm approaches standard 2D and 3D benchmark cases to demonstrate the process of these classic algorithms. It is noted for this chapter that the compliance minimization or stiffness maximization problem is considered as the objective function for all optimization cases. It is vital that the reader understands both the strengths and weaknesses of each algorithm since this will lead to greater comprehension of the Smooth-Edged Material Distribution for Optimizing Topology (SEMDOT) method — described in Chapter 3.

2.1 Introduction

This chapter will investigate how the classic topology optimization algorithms approach standard 2D and 3D cases; this will explain the underlying steps of each algorithm and the potential strengths and weaknesses of each algorithm. Figure 2.1 shows the standard 2D-loaded cantilever problem that will be analyzed by each topology optimization algorithm. The 2D-loaded cantilever beam has a design domain, D, where material can be placed or removed to support the load and to provide connection to the boundary constraint. The load, F, is attached to the bottom right corner of the design domain. The portion of the surface of the design domain connected to the load is given as, Γ_L. The left-hand edge of the design domain is fixed. The portion of the surface of the design domain connected to the boundary constraint (fixed wall) is given as Γ_C. Figure 2.1 shows a partially optimized cantilever beam. This figure shows the nomenclature for the optimized cantilever beam, such that the internal solid domain represented by Ω, the void domain represented by $D\backslash\Omega$, and the boundaries between the solid and void domains represented by $\delta\Omega$ can be defined. The 2D case to be optimized in this chapter is shown in Figure 2.2. The 3D case is also a cantilever beam, except the load is applied to the center of the free end surface of the beam, see Figure 2.3.

In very simple terms, the first two classic topology optimization algorithms (SIMP and BESO) are density-based algorithms that break up the design domain, D, into small regular elements, where the amount of material in each element is defined by the local density. Finite Element Analysis (FEA) can then be conducted to determine the stress in each element. The schemes then apply different methods to rank the importance of the element to the structure, and the insignificant elements will have material removed by decreasing the local density. The second two schemes (LSMs and MMC) define the boundaries of the solid part, $\delta\Omega$, or the geometric features that make up the part. The boundaries or features are moved and scaled by the optimization algorithms to minimize the mass of the solid part while maintaining the constraints and the connection to the loads.

DOI: 10.1201/9781032634449-2

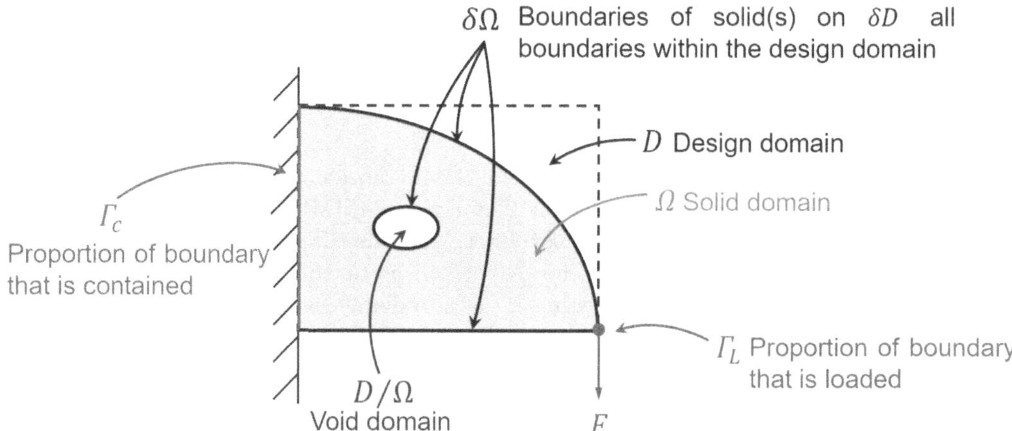

FIGURE 2.1
Nomenclature of a basic topology optimization case, where D is the design domain, Γ_C is the portion of the boundary that is constrained, Γ_L is the portion of the boundary that is loaded, Ω is the solid domain of the beam, $D\backslash\Omega$ is the void domain, and $\delta\Omega$ is the boundary of the solid(s) of the beam.

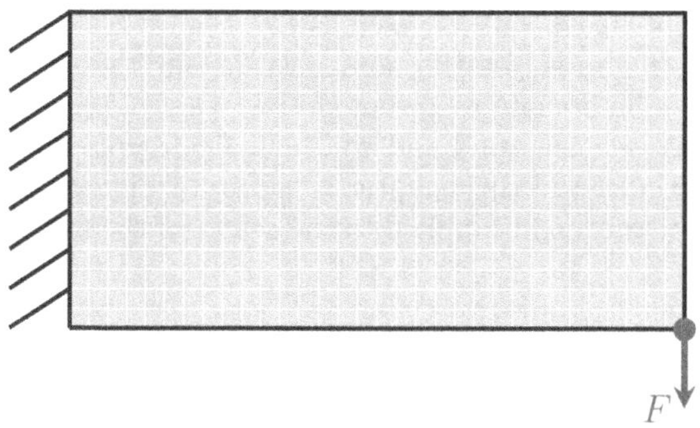

FIGURE 2.2
The 2D cantilever beam case.

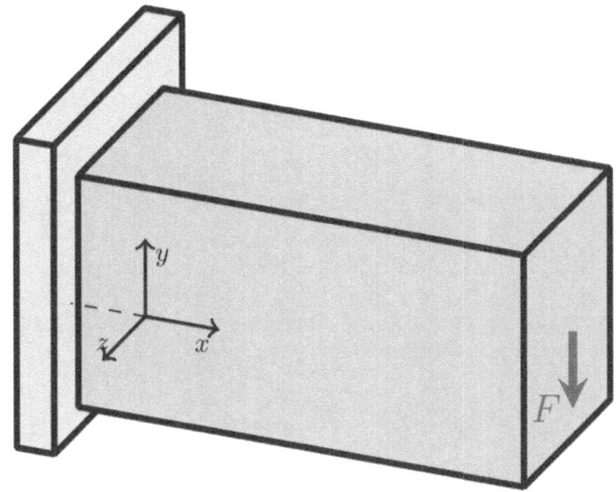

FIGURE 2.3
The 3D cantilever beam case.

2.2 Solid Isotropic Material with Penalization (SIMP) Method

SIMP is the first classic topology optimization algorithm. It is defined as a density-based optimization algorithm because it minimizes the mass by modifying the local density. Density-based topology optimization methods use a fixed design domain, D, of finite elements to achieve a basic mechanical performance goal given some boundary and loading conditions. In detail, D is broken into a set of finite elements where each element has a local density. The SIMP optimization algorithm then suggests a material distribution across D to satisfy all the constraints and loads in the system. Penalization of each element's density is used in the optimization method to push each element in the design domain toward the state of being solid or void [37]. Undoubtedly, the most popular and important density-based approach is the SIMP method [20]. The SIMP method was originally devised by Bendsøe [17] and Zhou and Rozvany [218] independently, and the early definition of term "SIMP" can be found in [153]. It should be mentioned that the publication of the 99-line MATLAB code by Sigmund [158] greatly contributed to the ubiquitous use of SIMP and rapid development of topology optimization. Open source codes of SIMP can be found in [1, 10, 43, 118, 158]. In addition, SIMP is the standard topology optimization algorithm provided within many commercial FEA software packages, such as ANSYS, Abaqus, Optistruct, and LS-TaSC.

2.2.1 Algorithm Framework

SIMP uses continuous design variables, where the density value will vary between 0 (void) and 1 (solid). As the SIMP optimization progresses, the intermediate elements that have a density value between 0 and 1 will be inevitably generated. The intermediate elements are not desired since they represent a wide range of material properties, and realistically they have only a small effect on the overall stiffness of the structure. Generally, the material penalization scheme is used to suppress intermediate elements by pushing these elements toward "0" or "1", or in other words, solid or void with the aim of producing the optimal design [167]. The basic framework of SIMP is as follows:

1. Make an initial design system with D, boundary constraints, and loadings

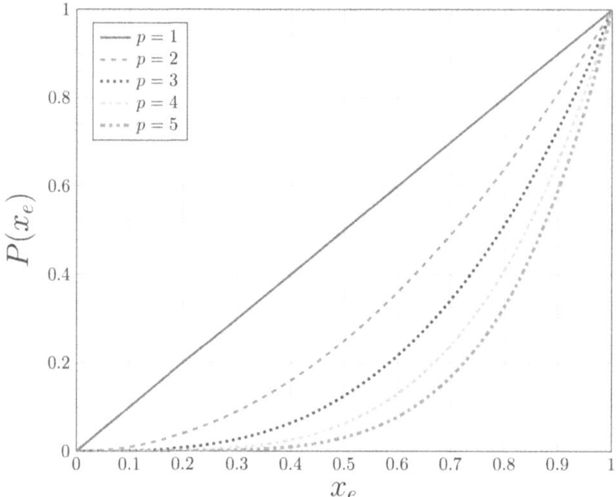

FIGURE 2.4
Penalization scheme in SIMP.

2. Mesh D
3. Start loop
   (a) Compute displacements and strains in the system;
   (b) Compute the compliance of the system;
   (c) Update the density variables based on the sensitivity analysis.
4. Repeat loop until converged

The penalization scheme used in SIMP is as follows:

$$P(x_e) = x_e^p, \quad x_e \in [0, 1], \tag{2.1}$$

where x_e is the density variable, p is the penalty coefficient, and $P(x_e)$ is the power-law function with respect to x_e. The SIMP penalization scheme is depicted in Figure 2.4. When using $p = 1$, there is no penalization of the material. In this case it is still a convex optimization problem with a unique solution of the material distribution in the design domain; however, the solution has a wide range of densities in the design domain, or in other words there are a range of material properties across the elements [144]. When using SIMP, $p > 1$ is used to penalize intermediate densities to produce solutions with a very high majority solid or void elements. The readers should understand that a very low value of p above 1.0 will generate too many "gray" elements, and a very high value of p will result in a too fast convergence to a local minima solution. The "magic number" for the penalty coefficient is said to be $p = 3$, which is able to ensure good convergence and almost total solid or void element solutions, and hence $p = 3$ is the default penalty coefficient value in SIMP.

Instead of using a fixed penalty coefficient, $p = 3$, a continuation method can be used on the penalty coefficient p to both penalize intermediate elements and mitigate the premature convergence to a local minima. The continuation method uses a function of the design variable to provide a more efficient optimization. For example, the strategy proposed by Groenwold and Etman [65] is expressed by:

$$p_k = \begin{cases} p_0 & k \leq k_{p_0}, \\ \min\{p_{\max}, \psi p_{k-1}\} & k > k_{p_0}, \end{cases} \tag{2.2}$$

where k is the current iteration, p_0 is the initial penalty factor, $p_{\max}$ is the maximum penalty factor, k_{p_0} is the number of iterations for p_0, and ψ is the parameter that controls the increase rate of p. Other than Equation (2.2), more continuation methods on p can be found in [42,54]. The readers should note that proper selection of the continuation strategy is based on experience or trial-and-error attempts. Generally, a proper continuation approach on p will lead to improved results.

The relationship between the density variable x_e and elemental stiffness modulus $E_e(x_e)$ is given by:

$$E_e(x_e) = x_e^p E_0, \quad x_e \in [0, 1], \tag{2.3}$$

where E_0 is the elastic Young's modulus of the solid material. Elements with zero stiffness may cause instabilities in the finite element analysis. In Equation (2.3), elements with zero stiffness can be avoided through adopting a lower stiffness limit that is marginally larger than zero on the density variables x_e.

After computing the compliance of the system, the sensitivity analysis must be completed. Based on Equation (2.3), sensitivities of the objective function c with respect to x_e are calculated by:

$$\frac{\partial c}{\partial x_e} = -p x_e^{p-1} \mathbf{u}_e^{\mathrm{T}} \mathbf{k}_e \mathbf{u}_e, \tag{2.4}$$

where $\mathbf{u}_e$ is the elemental displacement vector and $\mathbf{k}_e$ is the elemental stiffness matrix.

A modified relationship between x_e and $E_e(x_e)$ is:

$$E_e(x_e) = E_{\min} + x_e^p (E_0 - E_{\min}), \quad x_e \in [0, 1], \tag{2.5}$$

where $E_{\min}$ is the elastic modulus of the void material, which is a very small (non-zero) value, for example, 1×10^{-9}, to avoid a singularity of the elemental stiffness matrix. Compared to Equation (2.3), Equation (2.5) interpolates between the minimum elastic modulus value and penalization power and allows for a straightforward implementation of additional filters that will be introduced later [159].

Based on Equation (2.5), sensitivities of c with respect to x_e are calculated by:

$$\frac{\partial c}{\partial x_e} = -p x_e^{p-1} (E_0 - E_{\min}) \mathbf{u}_e^{\mathrm{T}} \mathbf{k}_0 \mathbf{u}_e, \tag{2.6}$$

where $\mathbf{k}_0$ is the stiffness matrix for an element with the unit Young's modulus.

To ensure the existence of solutions in topology optimization and to avoid the checkerboard pattern problem (see Figure 2.5), various filtering techniques are used to impose some restriction on the design [164]. Two representative filtering techniques that are extensively used in topology optimization are sensitivity and density filters.

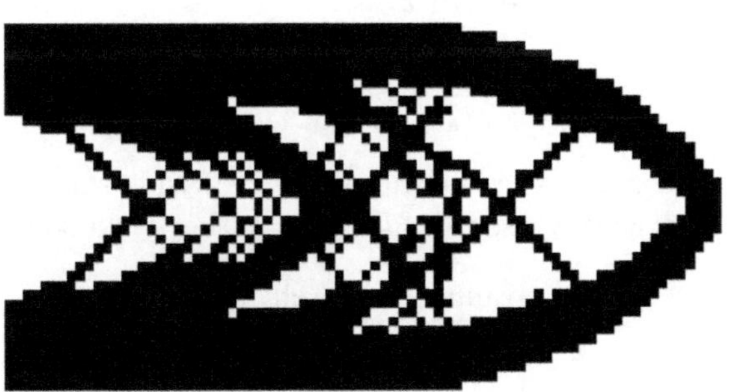

FIGURE 2.5
An example of checkerboard pattern.

The sensitivity filter modifies the sensitivities $\partial c/\partial x_e$ in the following form:

$$\frac{\hat{\partial c}}{\partial x_e} = \frac{1}{\max(\gamma, x_e) \sum_{i \in N_e} H_{ei}} \sum_{i \in N_e} H_{ei} x_i \frac{\partial c}{\partial x_i}, \tag{2.7}$$

where γ is a small positive number (for example, 0.001) to avoid division by zero, N_e is the set of elements i for which the center-to center distance $\Delta(e,i)$ to element e is smaller than the filter radius $r_{\min}$. The weight factor H_{ei} is defined by:

$$H_{ei} = \max(0, r_{\min} - \Delta(e,i)). \tag{2.8}$$

The reader should understand that the filter radius is determined by experience.

The density filter transforms the original density variables x_e in the form:

$$\tilde{x}_e = \frac{1}{\sum_{i \in N_e} H_{ei}} \sum_{i \in N_e} H_{ei} x_i, \tag{2.9}$$

where $\tilde{x}_e$ represents the filtered or physical densities. It is noted that the filtered density field $\tilde{x}_e$ should always be used as the solution to the optimization problem since the original densities x_e lose their physical meaning after using the density filter.

Provided that x_e is replaced with $\tilde{x}_e$, sensitivities with respect to the design variables x_j are calculated based on the chain rule:

$$\frac{\partial c}{\partial x_j} = \sum_{e \in N_j} \frac{\partial c}{\partial \tilde{x}_e} \frac{\partial \tilde{x}_e}{\partial x_j} = \sum_{e \in N_j} \frac{1}{\sum_{i \in N_e} H_{ei}} H_{je} \frac{\partial c}{\partial \tilde{x}_e}. \tag{2.10}$$

The Heaviside projection filter, a modification of the density filter in Equation (2.9), is generally used to further obtain solid/void solutions [66,159], which has the form:

$$\bar{x}_e = 1 - e^{-\beta \tilde{x}_e} + \tilde{x}_e e^{-\beta}, \tag{2.11}$$

where β is the parameter that controls the smoothness of the approximation. A continuation method is generally used to increase β gradually.

Sensitivities of c with respect to $\tilde{x}_e$ are obtained using the chain rule:

$$\frac{\partial c}{\partial \tilde{x}_e} = \frac{\partial c}{\partial \bar{x}_e} \frac{\partial \bar{x}_e}{\partial \tilde{x}_e}. \tag{2.12}$$

The derivative of $\bar{x}_e$ with respect to $\tilde{x}_e$ is:

$$\frac{\partial \bar{x}_e}{\partial \tilde{x}_e} = \beta e^{-\beta \tilde{x}_e} + e^{-\beta}. \tag{2.13}$$

Other than the Heaviside projection filter, the Heaviside smooth function proposed by Wang et al. [180] is extensively used in SIMP to further yield solid/void solutions, which has the form:

$$\bar{x}_e = \frac{\tanh(\beta\eta) + \tanh(\beta(\tilde{x}_e - \eta))}{\tanh(\beta\eta) + \tanh(\beta(1-\eta))}, \tag{2.14}$$

where η is the threshold value between 0 and 1 and is adjusted to satisfy the volume constraint during optimization. The Heaviside smooth function with $\eta = 0.5$ is illustrated in Figure 2.6.

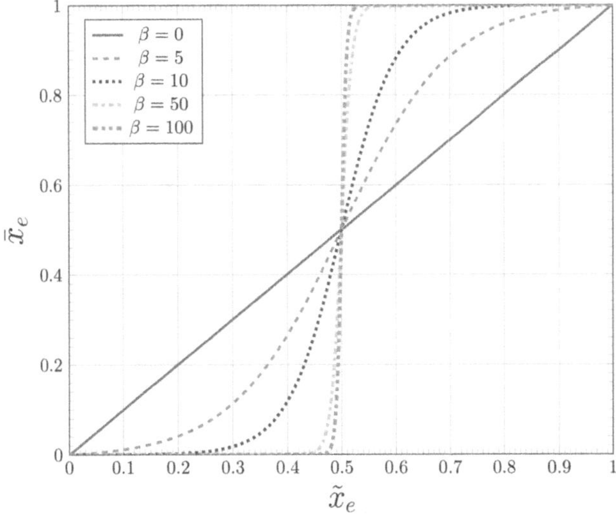

FIGURE 2.6
Heaviside smooth function.

After using Heaviside filters, the discreteness of resulting topologies can be measured by a gray level indicator:

$$M_{nd} = \frac{\sum_{e=1}^{n} 4\bar{x}_e(1 - \bar{x}_e)}{n} \times 100\%, \tag{2.15}$$

where n is the total number of elements. If all the elements are solid/void, $M_{nd} = 0\%$, and if all the elements are intermediate, $M_{nd} = 100\%$.

To terminate the optimization procedure in SIMP, the maximum variation of the design variable should be less than a predefined value, for example, 0.01. The termination criterion is therefore defined as:

$$\max(|x_e^k - x_e^{k-1}|) \leq 0.01. \tag{2.16}$$

2.2.2 Numerical Examples

Here, the 2D and 3D cantilever beam optimization problems based on Figures 2.1–2.3 are used to demonstrate the algorithm of SIMP. For simplicity in both the 2D and 3D examples, the Young's modulus is set to 1, and the Poisson's ratio is set to 0.3. Different filters (that is, sensitivity, density, and Heaviside projection filters) are taken into account in the 2D case, and the density filter is used in the 3D case. The design domain and boundary condition of the 2D cantilever beam case are shown in Figure 2.2. A 100×50 mesh is used, and the target volume is set to 0.3. For different filters, the filter radius $r_{\min}$ is set to 2 element widths.

The optimization convergence processes and the resultant topologies obtained by SIMP with a sensitivity filter, a density filter, and a Heaviside projection filter are shown in Figure 2.7–2.9, respectively. The use of a particular filter will lead to different distinct topological configurations.

The design domain and boundary condition of the 3D cantilever beam case are shown in Figure 2.3. A 60×30×20 mesh is used; the filter radius $r_{\min}$ is again set to 2 element widths; and the target volume is set to 0.1. The convergence process and resulting topology obtained by SIMP with the density filter for the 3D case are shown in Figure 2.10.

It can be seen in Figures 2.7(a), 2.8(a), 2.9(a), and 2.10(a) that the convergence processes of SIMP with the different filters are monotonic, and the objective function decreases from

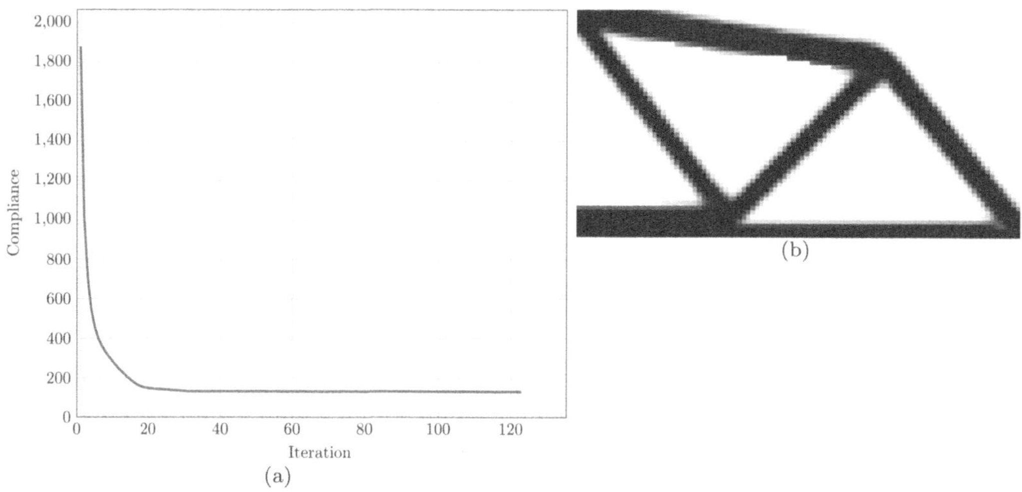

FIGURE 2.7
A 2D example of SIMP with sensitivity filter: (a) convergence process and (b) resulting topology.

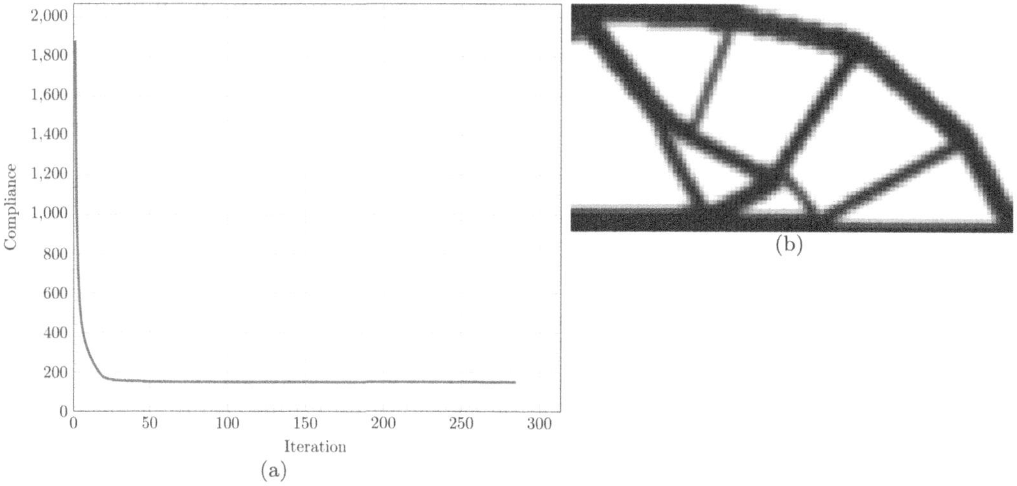

FIGURE 2.8
A 2D example of SIMP with density filter: (a) convergence process and (b) resulting topology.

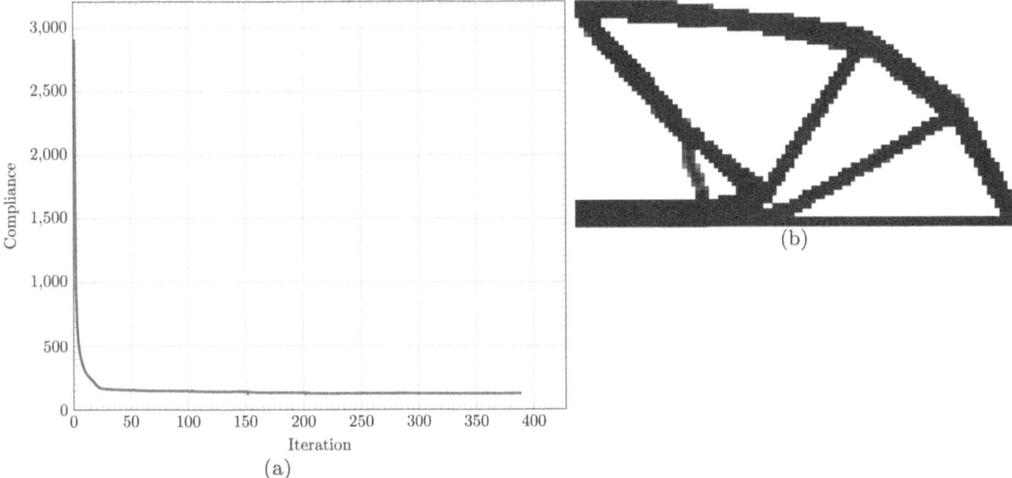

FIGURE 2.9
A 2D example of SIMP with Heaviside projection filter: (a) convergence process and (b) resulting topology.

a high value to an almost constant value. When using sensitivity and density filters, gray elements can be clearly observed in resulting topologies (see Figures 2.7(b), 2.8(b), and 2.10(b)). By contrast, the use of the Heaviside projection filter can yield a nearly solid-void solution (see Figure 2.9(b)).

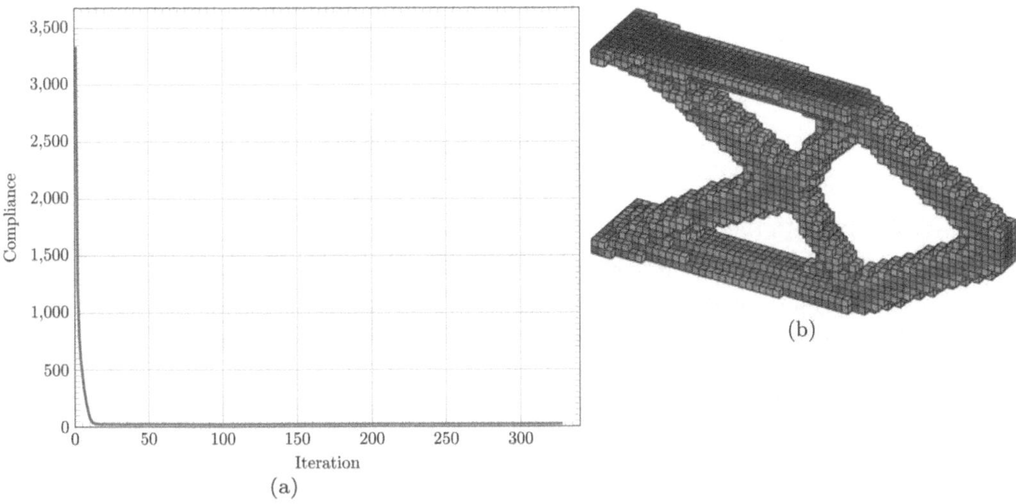

FIGURE 2.10
A 3D example of SIMP with density filter: (a) convergence process and (b) resulting topology.

2.3 Bi-directional Evolutionary Structural Optimization (BESO) Method

BESO algorithm, proposed by Huang and Xie [79,80], is the most popular discrete density-based approach. As a matter of fact, BESO is an advancement of the Evolutionary Structural Optimization (ESO) algorithm that was first introduced by Xie and Steven [197]. ESO incrementally erodes away the material in the design domain that has little influence on the structure's stiffness. The extension to BESO included an additional step of providing more material to the design domain where the material has a strong influence on the structure. BESO can be grouped into hard-kill and soft-kill approaches based on whether the void element is completely removed from the design domain or assigned a very small value [80,81]. This chapter only focuses on the modern version of BESO, which is the soft-kill approach. BESO solves the optimization problem using discrete design variables ($x_{\min}$ or 1). As explained above, during the optimization process BESO adds efficient and influential material to the structure and removes the inefficient and insignificant material at each increment from the structure. Although Sigmund and Maute [163] stated that the BESO methods should be regarded as a type of discrete SIMP approach because they use the material penalization scheme, the originality and novelty of the BESO algorithm cannot be denied and, hence, it still should be regarded as a pioneering algorithm in its own regard. Open source codes of BESO can be found in [81,224]. In addition, the commercial software package named Ameba, developed by Prof. Yi-Min Xie's team, provides a topology optimization tool based on BESO.

2.3.1 Algorithm Framework

The basic framework for the BESO algorithm is as follows:

1. Make an initial design system with D, boundary constraints, and loadings
2. Mesh D
3. Start loop
   (a) Compute the displacements and strains;
   (b) Compute the compliance of the system;
   (c) Compute the element sensitivity with respect to the compliance;
   (d) Remove an Evolutionary Rate (ER) % of elements with sensitivities less than the reject ratio;
   (e) Add an ER % of solid elements around the elements with large sensitivities if the target volume criterion is not met.
4. Repeat the loop until converged

BESO also uses the penalization scheme to steer the optimization solution toward a separated solid-void optimal design. However, it is not possible to find the existence of a single convergent binary fully separated solid-void elemental design for any general structural optimization scenario. Nevertheless, Xia et al. [196] show that the evolutionary structural optimization methods can provide a robust binary structural solution. In BESO, the material binary description is as follows:

$$E_e(x_e) = x_e^p E_0, \quad x_e \in \{x_{\min}, 1\}, \tag{2.17}$$

where $x_{\min}$ is a small value for void elements, for example, 0.001. Here, x_e represents the binary design variables.

The elements in design volume must be ranked in terms of their influence on the stiffness of the overall structure. The influence is measured by the variation of the element compliance when a given element is added or removed. Sensitivities for solid and void elements are expressed by:

$$\alpha_e = -\frac{1}{p}\frac{\partial c}{\partial x_e} = \begin{cases} \mathbf{u}_e^{\mathrm{T}}\mathbf{k}_0\mathbf{u}_e & x_e = 1, \\ x_{\min}^{p-1}\mathbf{u}_e^{\mathrm{T}}\mathbf{k}_0\mathbf{u}_e & x_e = x_{\min}, \end{cases} \tag{2.18}$$

where α_e is defined as the sensitivity number for the eth element. The ER, defined by the user for the optimization, then defines how fast the low-sensitivity solid elements are eroded and how fast the high-sensitivity void elements are converted to solid elements. The readers should note that the ER is usually defined by experience of the user with regard to the problem, but commonly it is set at 2%.

To avoid mesh-dependency and checkerboard patterns, the sensitivity numbers are modified via a filter:

$$\hat{\alpha}_e = \frac{\sum_{n_t=1}^{N_t} \omega(r_{en_t})\alpha_{n_t}^n}{\sum_{n_t=1}^{N_t} \omega(r_{en_t})}, \tag{2.19}$$

where N_t is the total number of nodes within a circle of radius $r_{\min}$ from the center of the eth element, r_{en_t} is the distance between the center of the eth element and n_tth node, and $\alpha_{n_t}^n$ is the nodal sensitivity number. The weight factor $\omega(r_{en_t})$ is given by:

$$\omega(r_{en_t}) = \begin{cases} r_{\min} - r_{en_t} & r_{en_t} < r_{\min}, \\ 0 & r_{en_t} \geq r_{\min}. \end{cases} \tag{2.20}$$

The objective function and corresponding topology may not be convergent since the sensitivity numbers are calculated based on the particular status of the elements (1 or $x_{\min}$). The BESO method solves this issue by averaging the sensitivity number with its past number, which has the form:

$$\tilde{\alpha}_e = \frac{1}{2}(\hat{\alpha}_{e,k} + \hat{\alpha}_{e,k-1}), \tag{2.21}$$

where k is the iteration number.

To terminate the BESO optimization process, the predefined target volume V^* should be reached first, and then the following convergence criterion must be satisfied:

$$error = \frac{\left|\sum_{z=1}^{N_z} \left(c_{k-z+1} - c_{k-N_z-z+1}\right)\right|}{\sum_{z=1}^{N_z} c_{k-z+1}} \leq 0.001, \tag{2.22}$$

where N_z is the maximum number of iterations of compliance differentials progressing back from the current iteration. N_z is generally set to 5, which implies the termination criterion is true when the error is stable and less than the convergence threshold for at least 10 successive iterations.

In summary, the detailed optimization process of BESO is as follows:

1. Mesh the design domain using finite elements.

2. Define the parameters in BESO, such as the target volume V^*, evolutionary ratio ER, and penalty coefficient p.
3. Execute FEA.
4. Calculate the sensitivity numbers (Equation 2.18) and update sensitivity numbers by filtering (Equation 2.19) and averaging with past sensitivity information (Equation 2.21).
5. Determine V^* for the next iteration. If the current volume V_k is larger than V^*, the volume for the next iteration can be calculated by:

$$V_{k+1} = V_k(1 - ER). \tag{2.23}$$

   If the calculated volume for the next iteration is less than V^*, V_{k+1} is set to V^*.
6. Reset the design variables. For solid elements, the elemental density is switched from 1 to $x_{\min}$ if

$$\alpha_e \leq \alpha_{th}. \tag{2.24}$$

   For void elements, the elemental density is switched from $x_{\min}$ to 1 if

$$\alpha_e > \alpha_{th}, \tag{2.25}$$

   where α_{th} is the threshold of sensitivity numbers, which can be determined by the target volume and ranking of sensitivity numbers.
7. Repeat steps 3 to 6 until V^* is reached, and the termination criterion in Equation (2.22) is met.

2.3.2 Numerical Examples

Here, the 2D and 3D cantilever beam optimization problems based on Figures 2.1–2.3 are used to demonstrate the algorithm of BESO. For this example of BESO, the ER is set to 2% or 0.02. The optimization convergence process and resulting topology obtained by BESO for the 2D case are shown in Figure 2.11. The optimization convergence process and resulting

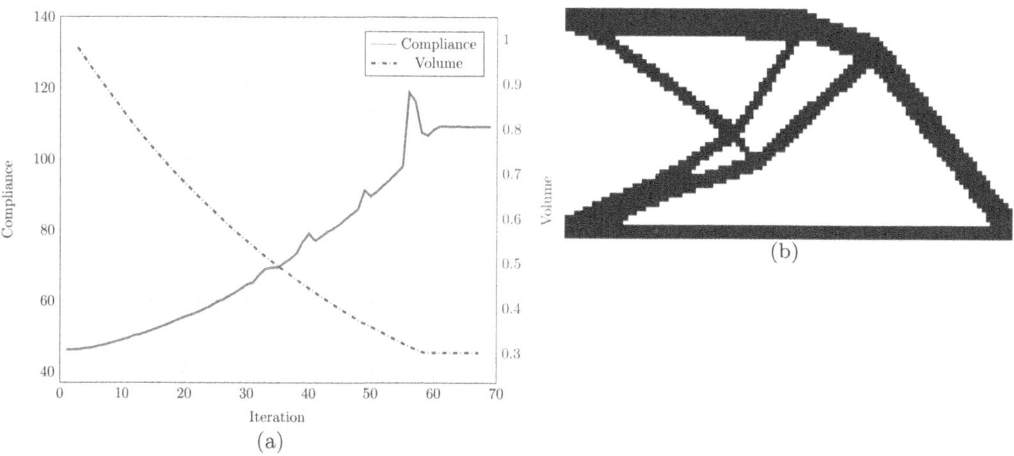

FIGURE 2.11
A 2D example of BESO: (a) convergence process and (b) resulting topology.

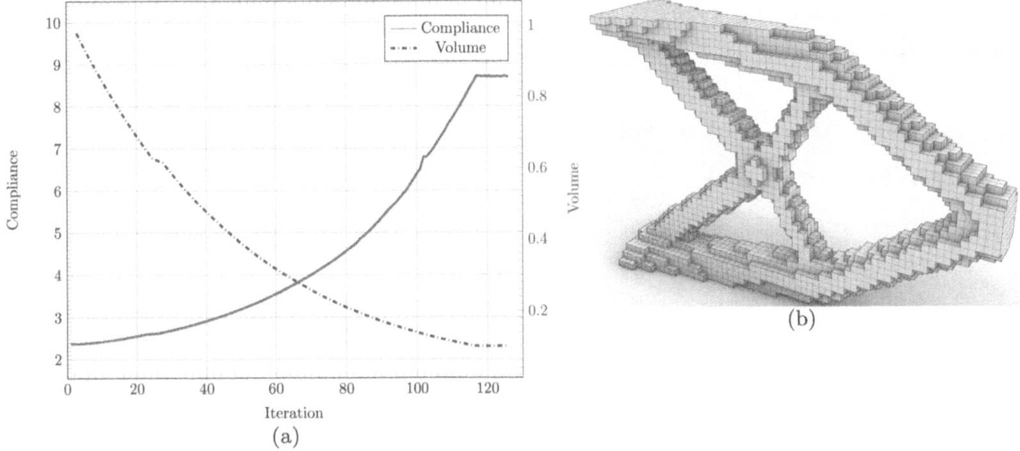

FIGURE 2.12
A 3D example of BESO: (a) convergence process and (b) resulting topology.

topology obtained by BESO for the 3D case are shown in Figure 2.12. Figures 2.11(a) and 2.12(a) show that the objective function (compliance) increases while the volume reduces according to Equation (2.23) at each iteration. Once the predefined target volume V^* is reached, then the volume converges to an almost constant value. The solid-void solutions using BESO can be observed in Figures 2.11(b) and 2.12(b).

2.4 Level-Set Method

LSMs use the iso-contours of a level-set function to mathematically define the interfaces between different materials (typically solid and void), and these interfaces can be used to describe the topological boundaries in the system [175]. The majority of LSMs are a generalized form of shape optimization with the potential to alter the topological configuration of the system by moving or generating boundaries. The initial concept of LSMs was proposed by Osher and Sethian [140], and the use of a level-set-based description for topology optimization was suggested by Haber and Bendsøe [68]. This was extended to use the LSM for structural optimization in the early works by Wang et al. [183] and Allaire et al. [8]. Essentially these methods are solving the Hamilton-Jacobi equation as applied to topology optimization systems to determine the boundary "velocity" field for the boundary update. LSMs can be categorized with respect to: the level-set function parameterization, the geometry mapping of the system, the mechanical model, the sensitivity function of the system, the update procedure, and finally, the applied regularization to provide sensitivity smoothing [175]. Results obtained by most LSMs do not encounter mesh-dependent spatial oscillations of the interface geometry that often occur in density-based methods; however LSMs suffer from local minima and a dependency on the initial starting configuration of the system. In addition, the results of LSMs heavily depend on the regularization techniques that are used to smooth the sensitivity information and reduce rapid movements in the interface geometries. Compared to density-based methods, such as SIMP and BESO, it is more difficult to understand the connection between the physical meaning of the optimization and the mathematics of the LSM. Open source codes for LSMs can be found in [25, 141, 186, 188, 204].

2.4.1 Algorithm Framework

The basic framework for the LSM algorithm is as follows:

1. Make a level-set function $\phi(\mathbf{x})$ such that $\phi(\mathbf{x}) = 0 \forall \mathbf{x} \in \delta\Omega$ on the boundary of the initial design system.
2. Prepare for determining topological derivatives of the system and the boundary "velocity" function.
3. Start loop:
   (a) Conduct finite element analysis to determine the displacements, $\mathbf{u}$, of the system;
   (b) Compute the sensitivity — topological derivatives;
   (c) Check whether the convergence criteria has been met or max iterations;
   (d) Compute the Lagrange multiplier λ for volume constraint;
   (e) Update the level-set function $\phi(\mathbf{x})$ using the boundary "velocity" function.
4. Repeat until convergence or max iterations.

First, the level-set function, which defines the interface or boundary between the solid and void materials, must be defined. The level-set function can be described as follows:

$$\begin{cases} \phi(\mathbf{x}) > 0 & \forall \mathbf{x} \in D \setminus \Omega, \\ \phi(\mathbf{x}) = 0 & \forall \mathbf{x} \ \ \delta\Omega, \\ \phi(\mathbf{x}) < 0 & \forall \mathbf{x} \ \ \Omega \setminus \delta\Omega, \end{cases} \tag{2.26}$$

where $\phi(\mathbf{x}) > 0$ is outside the structure and represents void space, and $\phi(\mathbf{x}) < 0$ is inside the structure and represents the solid space of the structure within the design domain, D. The boundary between solid and void is $\phi(\mathbf{x}) = 0$, and the first order time derivative of this boundary condition is:

$$\frac{\partial \phi(\mathbf{x})}{\partial t} + \nabla\phi(\mathbf{x}) \cdot \Upsilon(\mathbf{x}) = 0. \tag{2.27}$$

This is the Hamilton-Jacobi equation that must be solved to move the boundaries of the system toward an optimum structure or series of structures, where the time derivative of the level-set boundary function is equal to the divergence of the current structure's nodes, and $\Upsilon(\mathbf{x})$ is the velocity function of the level-set surface.

The level-set formulation must be applied to topology optimization, where the compliance is minimized in the system for a given volume of the design domain. Typically, the compliance is defined as $J(\Omega)$, which is the objective function of the optimization, where Ω is any possible solid within the design domain D, that is, $\Omega \subset D$. The optimization can then be described as follows:

$$\begin{cases} \text{Minimize:} \ \ J(\Omega) = \int_\Omega f \cdot u \, dv + \int_{\Gamma_L} g \cdot u \, ds \\ = \int_\Omega E_0 e(u) \cdot e(u) \, dv, \\ \text{Subject to:} \ \ \int_\Omega dv - V^* \leq 0, \end{cases} \tag{2.28}$$

where f and g are the loads and tractions respectively, E_0 is the elastic stiffness or Young's modulus, $e(u)$ is the strain tensor, and V^* is the maximum constrained volume of the structure. There are a number of different methods to determine the velocity function, $\Upsilon(\mathbf{x})$, of the level-set surface, and this depends on the complexity of the algorithm to explain the basic movement of the current surface, the generation of voids or new surfaces in the solid domain, and other topology issues. These are explained in the literature [100–104, 141, 154, 183, 205]. In the original work by Wang et al. [183], the velocity function is determined by solving the following equation:

$$-\int_D (\beta(u, \phi(\mathbf{x}) + \lambda_+)\, \delta(\phi(\mathbf{x}))\Psi(\mathbf{x})dv = \int_D \mu^2(\mathbf{x})\Upsilon(\mathbf{x})\Psi(\mathbf{x})dv, \tag{2.29}$$

where $\Psi(\mathbf{x}) \in C^0(D)$ for any continuous function of Ψ, $\beta(u, \phi(\mathbf{x}))$ is the sensitivity function of the objective function, $J(u, \Psi)$, $\mu(\mathbf{x})$ is a non-zero weighting function within the design domain D, and λ_+ is the Lagrange multiplier for the volume constraint of the system. The direction of the movement from the velocity function is always in the normal direction to the solid boundary, $\delta\Omega$. It should be noted that the velocity function often needs to be smoothed to provide incremental changes to the level-set function.

2.4.2 Numerical Examples

The LSM algorithm used in these numerical examples uses a reaction-diffusion equation (RDE) — which is based on a time-evolution scheme that can help remove checkerboard patterns. The 2D and 3D cantilever beam cases shown in Figures 2.2 and 2.3 demonstrate the algorithm mechanism of the RDE-based level-set method. Details on the mathematical descriptions of the compliance minimization problem in the RDE-based level-set method can be found in [103]. Here, the body-fitted mesh adaption method proposed by Allaire et al. [7] is used only for the 2D case to accurately describe the topological changes via an exact mesh of the geometry. For both 2D and 3D cases, the diffusive coefficient R_τ is set to 4, and the fictitious time step parameter Δt is set to 0.5.

The convergence process of the 2D case is shown in Figure 2.13, and the resulting topologies and the body-fitted mesh of the final design are shown in Figure 2.14. The convergence process of the 3D case is shown in Figure 2.15, and the resulting topologies are shown in Figure 2.16. The final design under various views is shown in Figure 2.17. As

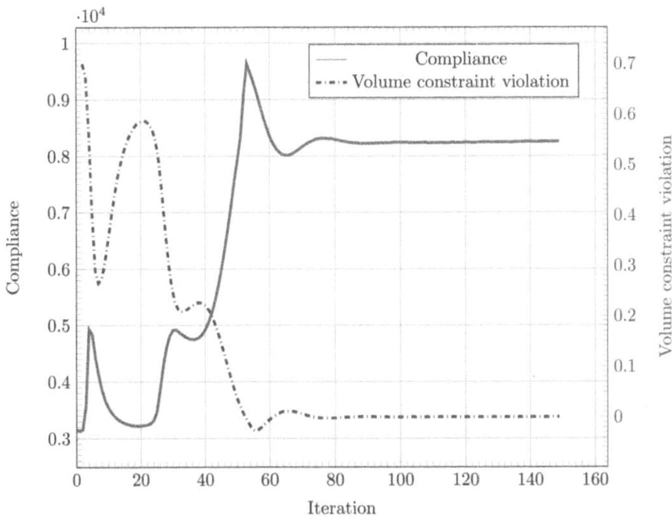

FIGURE 2.13
Convergence process of a 2D RDE example.

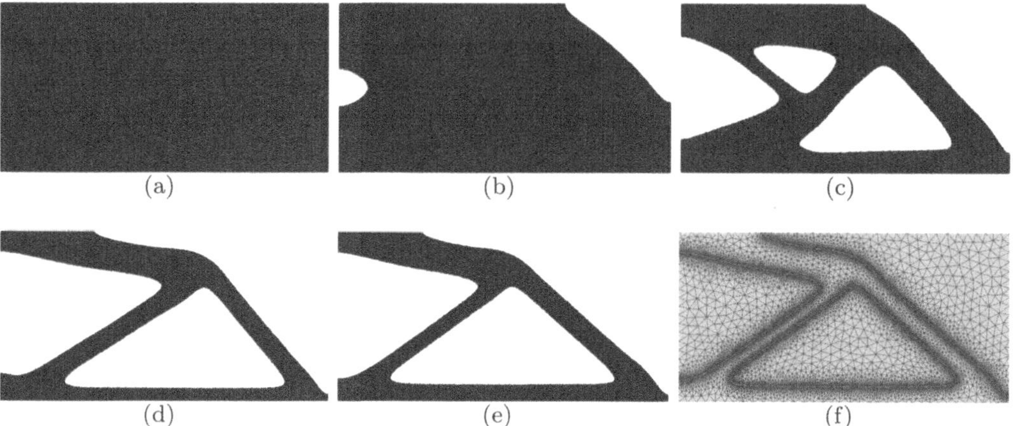

FIGURE 2.14
Resulting topologies for a 2D RDE example under (a) 0, (b) 20, (c) 40, (d) 60, and (e) 149 iterations, and (f) body-fitted mesh of final design.

shown in Figures 2.14 and 2.16, the RDE-based level-set method is able to generate smooth boundaries.

2.5 Moving Morphable Component (MMC) Based Method

The MMC-based approach is a feature-driven topology optimization method that was originally proposed by Guo et al. [67]. MMC implements topology optimization by manipulating geometrical features, otherwise known as morphable components, within the design domain in an explicit way. In MMC-based algorithms, all features are able to be translated, rotated, scaled, overlapped, and merged across the design domain, the topological configuration can

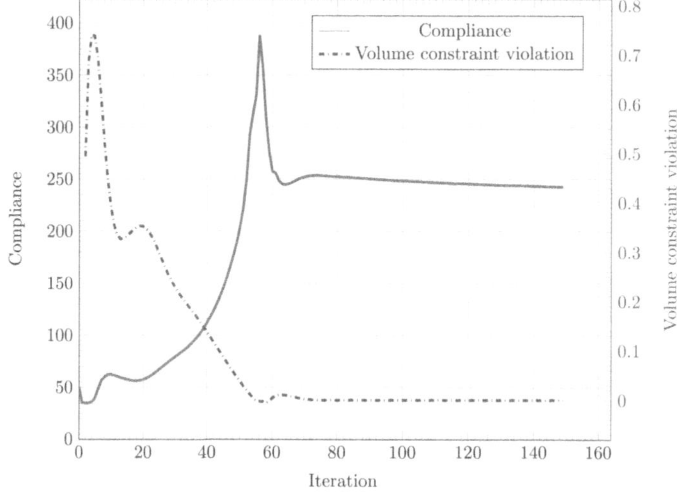

FIGURE 2.15
Convergence process of a 3D RDE example.

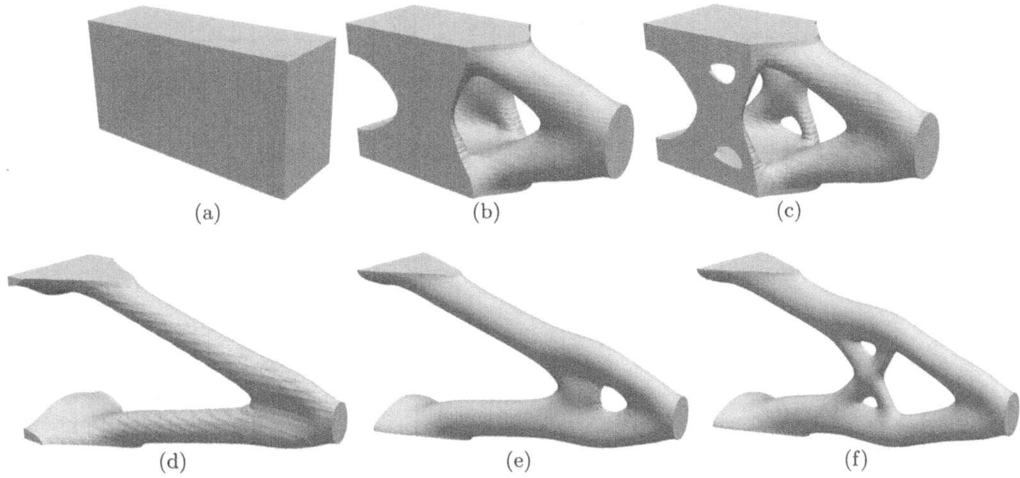

FIGURE 2.16
Resulting topologies for a 3D RDE example under (a) 0, (b) 20, (c) 30, (d) 60, (e) 80, and (f) 150 iterations.

be then obtained via optimizing the positions, rotation angles, and the scaled lengths and widths of the features [214]. The basic idea of the MMC-based method is illustrated in Figure 2.18. The MMC approach can identify the necessary number of features as well as the intersection relationships among the features based on the optimization of the geometric model. The number and shape of the geometric features are critical to the MMC optimization solution. Furthermore, local discontinuities inevitably appear at the overlap points of the features even when using sophisticated shapes to form boundaries. Smart algorithms are required to solve this issue. Open source codes of MMC-based approaches can be found in [39, 212].

2.5.1 Algorithm Framework

The basic framework of the MMC algorithm is as follows:

1. Define a Topology Description Function (TDF) with a design parameter change vector that can translate, rotate and scale a geometric feature in the design domain
2. Define a union function that merges all the geometric features in the MMC system
3. Start optimization loop

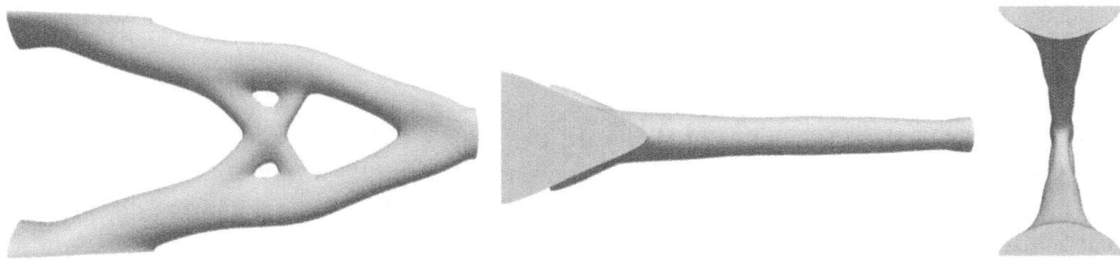

FIGURE 2.17
Resulting topology under different views.

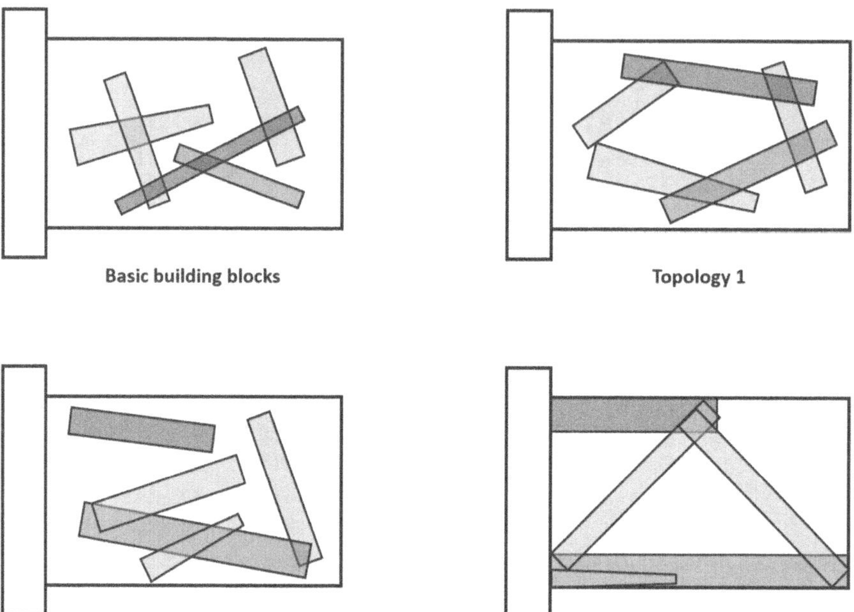

FIGURE 2.18
Basic idea of MMC-based approach.

Find the smallest series of design parameter changes in the MMC system that satisfy:

- The external load energy (based on body forces and surface tractions) equaling the internal energy (based on the stresses caused by the internal strain deformations);
- The volume of the MMC system, based on the union of all features, is less than the maximum volume, V^*;
- The system obeys the given boundary conditions;
- The system obeys the structural complexity constraints.

4. Repeat the optimization loop until the smallest series of design parameter changes in the MMC system

The first step of the MMC-based approach, based on [67], is to create a TDF that can translate, rotate, and scale a geometric feature in the design domain, D. The TDF for the ith feature Ω^i in the design domain, D, is defined by:

$$\begin{cases} \phi^i(x) > 0 & x \in \Omega^i, \\ \phi^i(x) = 0 & x \in \delta\Omega^i, \\ \phi^i(x) < 0 & x \in D \backslash \Omega^i, \end{cases} \tag{2.30}$$

where $\delta\Omega^i$ is the surface boundary of the ith feature.

The simplest application of the MMC approach is with 2D problems. For 2D cases, the geometry of the ith feature is defined by the following TDF, the reader should note that the feature is an isosceles trapezoid with parameters determined in [212]:

$$\phi^i = \left(\frac{x'}{L^i}\right)^{\varrho} + \left(\frac{y'}{l^i}\right)^{\varrho} - 1, \tag{2.31}$$

with

$$\begin{pmatrix} x' \\ y' \end{pmatrix} = \begin{bmatrix} \cos\theta_i & \sin\theta_i \\ -\sin\theta_i & \cos\theta_i \end{bmatrix} \begin{pmatrix} x - x_0^i \\ y - y_0^i \end{pmatrix}, \tag{2.32}$$

$$l^i = \frac{t_1^i + t_2^i}{2} + \frac{t_2^i - t_1^i}{2L^i} x', \tag{2.33}$$

where x' and y' are the coordinates of the center of the feature in the local coordinate system; L^i is the half-length of the feature; ϱ is a relatively large even integer number, which is generally set to 6; θ_i is the rotation angle of the feature; and x_0^s and y_0^s are the coordinates of the center of the feature in the global coordinate system. The readers should note that Equation (2.31) can be modified through exponentially scaling to ensure that the TDF values and their partial derivatives do not change too rapidly. Equations (2.32) and (2.33) are schematically illustrated in Figure 2.19.

For 3D cases, the geometry of the ith feature is represented by:

$$\phi^i = \left(\frac{x'}{L_1^i}\right)^{\varrho} + \left(\frac{y'}{L_2^i}\right)^{\varrho} + \left(\frac{z'}{L_3^i}\right)^{\varrho} - 1, \tag{2.34}$$

with

$$\begin{pmatrix} x' \\ y' \\ z' \end{pmatrix} = \begin{bmatrix} \cos B^i \cos\Theta^i & \cos B^i \sin\Theta^i & -\sin B^i \\ \sin A^i \sin B^s \cos\Theta^i - \cos A^i \sin\Theta^i & \sin A^i \sin B^i \sin\Theta^i + \cos A^i \cos\Theta^i & \sin A^i \cos B^i \\ \cos A^i \sin B^i \cos\Theta^i + \sin A^i \sin\Theta^i & \cos A^i \sin B^i \sin\Theta^i - \sin A^i \cos\Theta^i & \cos A^i \cos B^i \end{bmatrix} \begin{pmatrix} x - x_0^i \\ y - y_0^i \\ z - z_0^i \end{pmatrix}, \tag{2.35}$$

where x', y', and z' are the coordinates of the center of the ith feature in the local coordinate system; L_1^i, L_2^i, and L_3^i are the half-length, half-width, and half-height of the feature, respectively; A^i, B^i, and Θ^i are the rotation angles between the local and global coordinate systems; and x_0^i, y_0^i, and z_0^i are the coordinates of the center of the ith feature in the global coordinate system. Equation (2.35) can be schematically illustrated in Figure 2.20.

To obtain analytical sensitivities, the Kreisselmeier-Steinhauser (KS) function is used to approximate the union operation to approximate the overall structure. The KS function is defined as:

$$\phi^I \approx \left(\ln \left(\sum_{i=1}^{N} \exp\left(\lambda \phi^i\right) \right) \right) / \lambda, \tag{2.36}$$

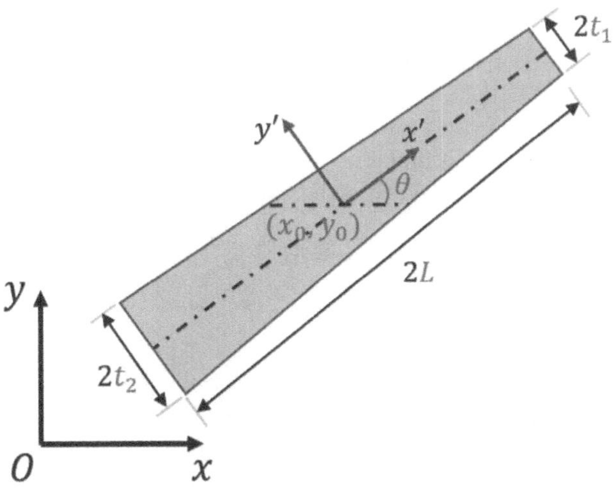

FIGURE 2.19
Illustration of a 2D feature in local and global Cartesian coordinate systems.

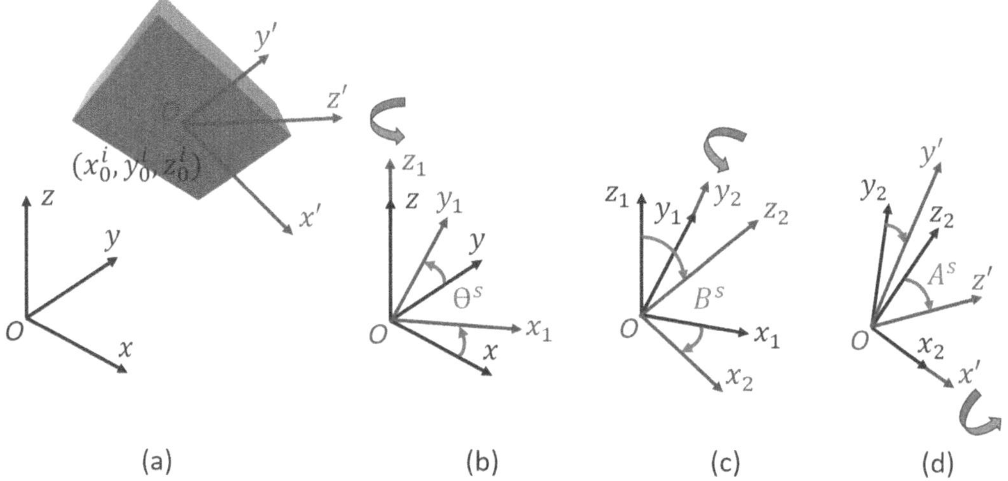

FIGURE 2.20
Illustration of a 3D feature in local and global Cartesian coordinate systems.

where N is the total number of features within the design domain, D, and λ is a large positive number, which is generally set to at least 100.

The Young's modulus is calculated by:

$$E_e = \rho_e E_0, \tag{2.37}$$

where ρ_e is the density of the eth element, which has the form:

$$\rho_e = \begin{cases} \frac{1}{4}\sum\limits_{i=1}^{4} H_\kappa\left(\phi_{e,s}^I\right) & \text{for 2D cases,} \\ \frac{1}{8}\sum\limits_{i=1}^{8} H_\kappa\left(\phi_{e,s}^I\right) & \text{for 3D cases,} \end{cases} \tag{2.38}$$

where $\phi_{e,s}^I$ is the value of the global TDF, ϕ^I, at the sth node of the eth element and $H_\kappa(\zeta)$ is the smoothed Heaviside function, which is defined as:

$$H_\kappa(\zeta) = \begin{cases} 1 & \zeta > \kappa, \\ \frac{3(1-\rho_{\min})}{4}\left(\frac{\zeta}{\kappa} - \frac{\zeta^3}{3\kappa^3}\right) + \frac{1+\rho_{\min}}{2} & |\zeta| \leq \kappa, \\ \rho_{\min} & \text{otherwise,} \end{cases} \tag{2.39}$$

where κ is the parameter controlling the size of the transition zone of H_κ and $\rho_{\min}$ is a small positive parameter introduced to avoid the possible singularity of the global stiffness matrix.

The volume constraint in MMC is given by:

$$g_v = V/V_D - \mathfrak{v} \leq 0, \tag{2.40}$$

where V is the volume of the current optimized structure, V_D is the volume of the design domain, and $\mathfrak{v} = \frac{V^*}{V_D}$ is the upper bound of the allowable volume fraction.

Sensitivities of the objective function with respect to each specific design variable d_j^i (the jth parameter associated with the ith component) are calculated by:

$$\frac{\partial c}{\partial d_\mathrm{j}^i} = \sum_{e=1}^{NE} \frac{\partial c}{\partial \rho_e}\frac{\partial \rho_e}{\partial d_\mathrm{j}^i} = -\sum_{e=1}^{NE} \mathrm{u}_e^{\mathrm{T}}\mathbf{k}_0\mathrm{u}_e \frac{\partial \rho_e}{\partial \phi^I}\frac{\partial \phi^I}{\partial \phi^i}\frac{\partial \phi^i}{\partial d_\mathrm{j}^i}, \tag{2.41}$$

where NE is the total number of elements in the structure.

Sensitivities of the volume constraint with respect to d_j^i are calculated by:

$$\frac{\partial g_v}{\partial d_j^i} = \sum_{e=1}^{NE} \frac{\partial g_v}{\partial \rho_e} \frac{\partial \rho_e}{\partial d_j^i} = \sum_{e=1}^{NE} \frac{V_e}{V_D} \frac{\partial \rho_e}{\partial \phi^I} \frac{\partial \phi^I}{\partial \phi^i} \frac{\partial \phi^s}{\partial d_j^i}, \tag{2.42}$$

where V_e is the volume of the eth element.

To terminate the optimization procedure in MMC, the relative variation of the objective function values in the last consecutive five iterations should be less than 1×10^{-4}. The relative variation is expressed by:

$$cVr5(iter) = \begin{cases} 1.0 & \text{if } iter < 5, \\ cVr5(iter-1) & \text{if } iter \geq 5 \text{ and } V_{er} > 1\times10^{-4}, \\ \left| \frac{\max\limits_{\mathfrak{L}=iter-4,\cdots,iter} \left| c(\mathfrak{L}) - \sum_{iter-4}^{iter} c(iter)/5 \right|}{\sum_{iter-4}^{iter} c(iter)/5} \right| & \text{else}, \end{cases} \tag{2.43}$$

where $V_{er} = (V/V_D - \mathfrak{v})/\mathfrak{v}$ with V representing the volume of the current optimized structure and $iter$ is the iteration number.

2.5.2 Numerical Examples

2D and 3D cantilever beam cases shown in Figures 2.2 and 2.3 are used to demonstrate the algorithm mechanism of MMC. In terms of the 2D case, the design domain has a size of 100×50 and is discretized by 100×50 elements. Here, $L = 30$, $t_1 = 4.5$, $t_2 = 4.5$, and $\theta = \text{asin}(0.7)$ are employed. The center coordinate of a representative component that is used to generate the initial design is (12.5, 12.5). In Equation (2.39), κ is set to 0.2, and ρ_{min} is set to 1×10^{-9}. The convergence process of the 2D case is shown in Figure 2.21, where the volume constraint violation is calculated by $V(D) - \mathfrak{v}$. When the optimization procedure converges, the volume constraint violation reaches a negligible value. Resulting topologies under different iterations are shown in Figure 2.22.

In the context of the 3D case, the design domain has a size of $60 \times 30 \times 20$ and is meshed by $60 \times 30 \times 20$ elements. Here, $L_1 = 12.5$, $L_2 = 2.5$, $L_3 = 1$, $A = 0$, $B = \text{atan}(0.6)$, and

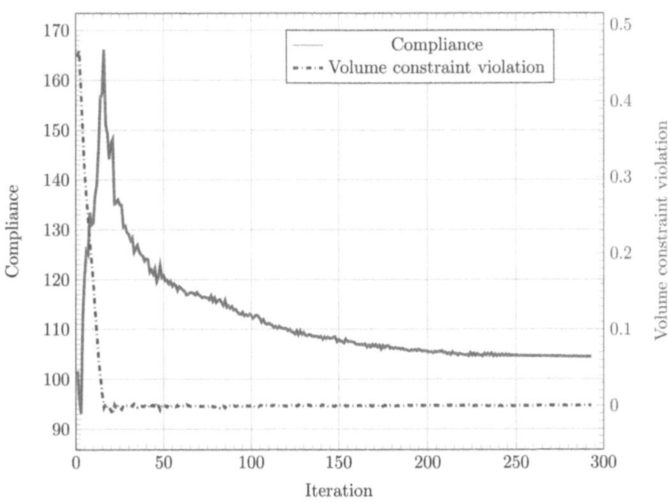

FIGURE 2.21
Convergence process of a 2D MMC example.

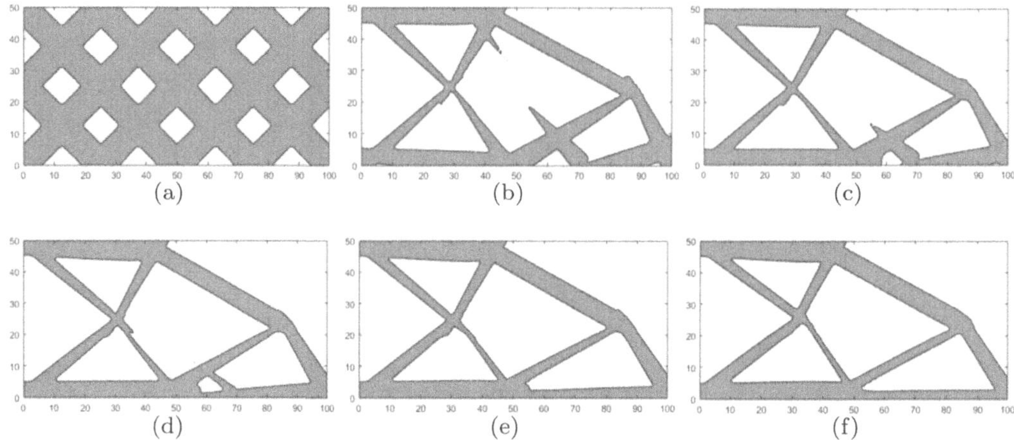

FIGURE 2.22
Resulting topologies for a 2D MMC example under (a) 0, (b) 50, (c) 100, (d) 150, (e) 200, and (f) 293 iterations.

$\Theta = 0$ are taken into account. The center coordinate of a representative component is (10, 10, 7.5). For the parameters related to Equation (2.39), κ is set to 0.2, and $\rho_{\min}$ is set to 1×10^{-3}. The convergence process of the 3D case is shown in Figure 2.23, and resulting topologies under different iterations are shown in Figure 2.24.

As can be seen in Figures 2.22(f) and 2.24(f), clear topological boundaries can be formed by MMC. Obviously, MMC can obtain smoother boundaries in comparison with SIMP and BESO.

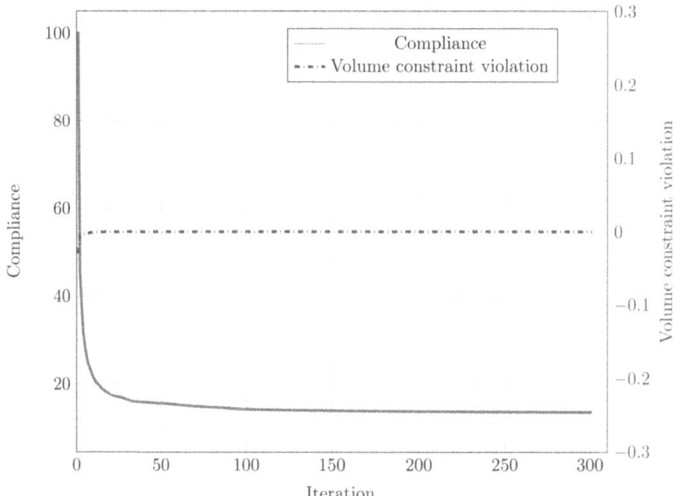

FIGURE 2.23
Convergence process of a 3D MMC example.

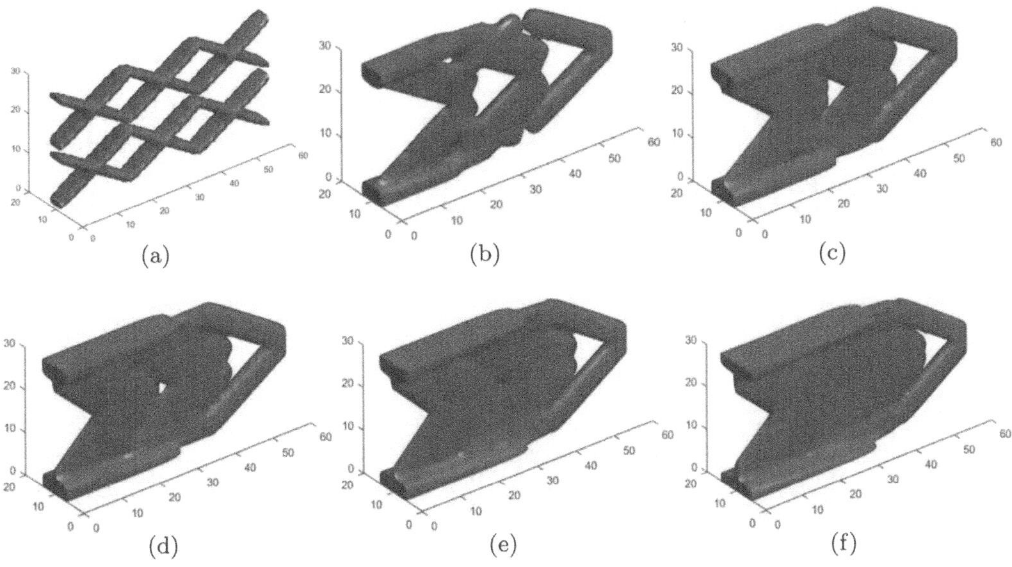

FIGURE 2.24
Resulting topologies for a 3D MMC example under (a) 0, (b) 50, (c) 100, (d) 150, (e) 200, and (f) 301 iterations.

2.6 Summary

This chapter comprehensively introduces the four classic representative topology optimization algorithms: SIMP, BESO, LSM-based, and MMC-based methods. The algorithm frameworks of these four methods are demonstrated in detail, and numerical examples of these four algorithms have been presented. The SIMP and BESO approaches are density-based algorithms, where the design domain, D, is broken into small finite elements and each element has a local density. SIMP uses the objective function sensitivity to scale the local densities across the design domain. BESO ranks the elements in the design domain by the objective function sensitivity, and less useful elements are incremental removed by minimizing the local density, while solid elements are added to regions of high sensitivity. The LSM-based and MMC-based approaches move boundaries or features to optimize the overall structure. LSM solves the Hamilton-Jacobi equation to move the boundaries of the solid structure in the design domain. MMC translates, rotates, and scales geometric features within the design domain to construct a union of features with an optimized topology. The creation of Smooth-Edged Material Distribution for Optimizing Topology (SEMDOT), in the next chapter, was motivated by the lack of smooth boundaries from the density-based methods and the lack of simplicity from the LSM and MMC methods.

3

Smooth-Edged Material Distribution for Optimizing Topology (SEMDOT) Algorithm

This chapter details the authors' and their colleagues new topology optimization algorithm named "Smooth-Edged Material Distribution for Optimizing Topology" (SEMDOT) [46, 47, 53, 54] based on the smooth boundary strategy presented in the Evolutionary Topology Optimization (ETO) method [33, 115], Huang's idea on smooth topological design [27], and the SIMP framework. The motivation behind the SEMDOT algorithm is comprehensively discussed based on the advantages and disadvantages of existing algorithms. This chapter focuses on the latest version of SEMDOT, which is a non-penalization topology optimization method that is an improvement on the original algorithm.

3.1 Motivation of Proposing a New Method

As discussed in Chapter 2, there are already a range of classic topology optimization algorithms and this begs the question why a new algorithm is still needed. Despite their easy implementation and ability to remove materials freely across the design domain, the element-based topology optimization algorithms will inevitably generate zigzag (discrete methods, such as BESO) or both zigzag and fuzzy boundaries (continuum methods, such as SIMP). Hence, post-processing smoothing methods are generally required to form smooth part surfaces after topology optimization to enable fabrication [109]. The post-processing methods not only require extra effort but also will increase the structural volume and mass when the maximum smooth part envelope is created from the topology optimization boundary [162]. Although level-set and MMC-based methods can generate smooth boundaries, they have some inherent disadvantages. The optimized results from level-set methods are highly dependent on the initial solid boundaries (for example, the number and location of holes) and the adopted regularization technique to ensure algorithm convergence. Moreover, in the MMC-based methods, there are issues of local non-smoothness that can appear at the intersection of the geometric features, even when using complicated feature shapes to obtain the boundaries. Compared to element-based approaches, the algorithm complexities of level-set and MMC-based methods are much higher (refer to Chapter 2). The high algorithm complexity may prevent these structural optimization tools from being used on complex problems. Therefore, element-based methods are generally preferred in solving complex and real-world structural optimization problems.

The use of the material penalization scheme in representative element-based approaches, such as SIMP and BESO, increases the non-linearity of the objective functional or constraint function, which may cause the optimization algorithm to easily run into local

DOI: 10.1201/9781032634449-3

optima [167]. The material penalization scheme may cause the overestimation of structural stiffness in density-based methods [193], and in terms of new optimization problems, the physical meaning of the material penalization model needs to be investigated and understood because of the nonlinear scaling of the material and its effect on multi-material structural design [75, 107]. In addition, the removal of the penalty coefficient from the material penalization scheme can further reduce the algorithm complexity.

Based on the above discussions, it can be concluded that there is a need from the topology optimization community for a new element-based approach that is capable of directly generating smooth topological boundaries and does not need to use the material penalization scheme. The remainder of this chapter will first discuss the details of the Smooth-Edged Material Distribution for Optimizing Topology (SEMDOT) algorithm, and this will be followed by a discourse on the sensitivity of SEMDOT's objective function. Finally, a flowchart is provided to breakdown the SEMDOT algorithm into its main components.

3.2 Algorithm Framework

The basic framework for the SEMDOT algorithm is as follows:

1. Start loop
   (a) Conduct finite element analysis
   (b) Conduct the sensitivity analysis with respect to the topological design variables
   (c) Update the densities at each element
   (d) Assign the density to the grid points for each element in the design volume
   (e) Construct the level-set and reassign the densities at the grid points
2. Repeat until converged or maximum iterations

The aims of proposing a new element-based algorithm named SEMDOT are to directly generate smooth topological boundaries and to remove the need for a material penalization scheme from topology optimization. The development of SEMDOT takes full advantage of the benefits of the smooth representation in the Evolutionary Topology Optimization (ETO) algorithm and element-based optimization in the SIMP algorithm. SEMDOT operates at two levels: elements and grid points. At the element level, an element-based approach is used to perform the main optimization using finite element analysis. This approach is similar to SIMP. However, associated with each element in the design volume is a matrix of grid points. At the grid-point level, the grid point values represent the material density distribution within the element. The grid point can take on two types of densities (solid or void). SEMDOT uses the level-set method to generate a series of smooth topological boundaries based on the solid/void interfaces at the grid-point level when combining all the grid point information from all the elements, as illustrated in Figure 3.1(a). The grid points are then assigned densities based on the solid/void interface boundaries within the design volume. The grid points that are contained within the solid region are assigned a solid density, and the grid points outside the solid region are assigned the artificial minimum density. Unlike the SIMP algorithm, which pursues the separation between solid and void elements, the basic idea of SEMDOT is to pursue the minimization of the perimeter (at the grid-point level) and width of the solid/void boundary (at the element level), where the width of the boundary represents the uncertainty of the optimized shape. Element properties are determined by whether the element's grid points are on, inside, or outside the

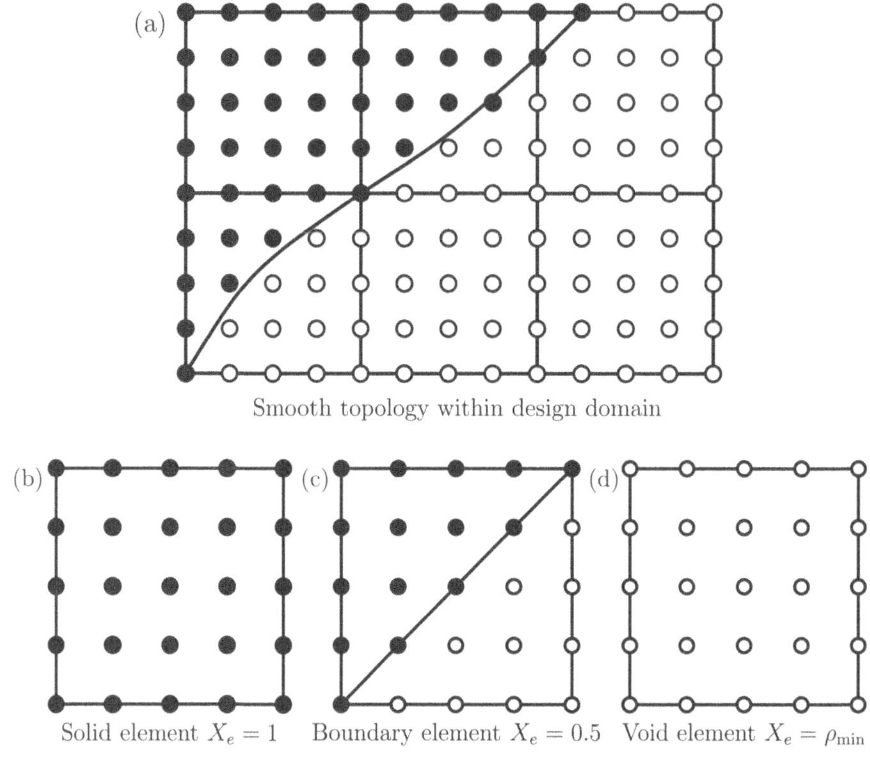

FIGURE 3.1
Illustration of smooth topology, solid, void, and boundary elements for 2D cases.

solid/void boundary; an element can then either be a solid, boundary/intermediate density or void. The elements that are fully within the boundary are solid, elements that are on the boundary have an intermediate density, and elements that are outside the boundary are void elements — and are assigned the minimum density, this is shown in Figure 3.1(b-d). It is noted that Figure 3.1 is the demonstration for 2D cases. For 3D cases, the smooth boundary strategy in SEMDOT is illustrated in Figure 3.2.

The grid point stiffness modulus of a grid point can be expressed by:

$$E_e(\rho_{e,g}) = \rho_{e,g}E^1, \ \rho_{e,g} \in \{\rho_{\min}, 1\}, \tag{3.1}$$

where $E_e(\rho_{e,g})$ is the function of the grid point stiffness modulus with respect to grid point densities, E^1 is the Young's modulus of the solid material, $\rho_{e,g}$ is the scaled value of the gth grid point assigned to the eth element, which is named the grid point density from here on (see Figure 3.1) rather than the physical density, and $\rho_{\min}$ is a small artificial value to represent the void density (for example, 0.001). It is noted that $\rho_{e,g}$ is assigned 1 to represent a solid grid point and $\rho_{\min}$ to represent a void grid point (see Figure 3.1).

SEMDOT uses the SIMP framework at the element level; therefore, element-based variables are needed in SEMDOT to conduct the finite element analysis. Elemental volume fractions that depend on grid point densities are defined as surrogate design variables such that:

$$X_e = \frac{1}{N}\sum_{g=1}^{N} \rho_{e,g}, \tag{3.2}$$

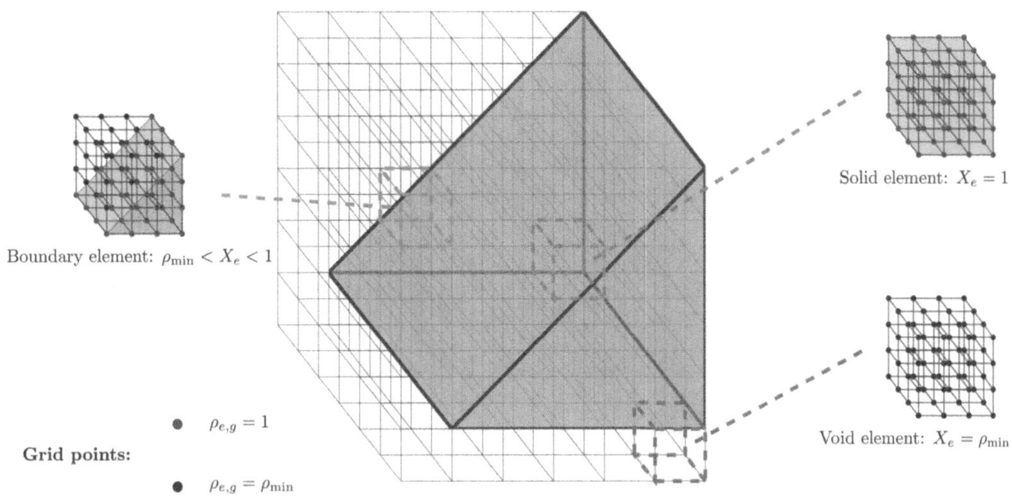

FIGURE 3.2
Illustration of smooth topology, solid, void, and boundary elements for 3D cases.

where X_e is the volume fraction of the eth element and N is the total number of grid points in each element. The function of Equation (3.2) is to convert discrete variables (grid point densities $\rho_{e,g} \in \{\rho_{\min}, 1\}$) to continuous variables (elemental volume fractions $X_e \in [\rho_{\min}, 1]$).

As shown in Figure 3.1(b, d), grid point densities are homogeneously distributed across the solid (with a relative value of $X_e = 1$) or void (with a relative value of $X_e = \rho_{\min}$) elements. Based on Equation (3.2), the elemental stiffness modulus of a solid or void element is expressed by:

$$E_e(X_e) = X_e E^1, \ X_e \in [\rho_{\min}, 1]. \tag{3.3}$$

Using the same form, the stiffness matrix of a solid or void element is:

$$\mathbf{K}_e(X_e) = X_e \mathbf{K}_e^1, \tag{3.4}$$

where $\mathbf{K}_e^1$ is the stiffness matrix of the solid element.

The main difference between SIMP and SEMDOT is the boundary/intermediate elements. In SIMP, all elements have intermediate density values between 0 and 1. In SEMDOT, only the boundary elements are intermediate with element density values between $\rho_{\min}$ and 1. The boundary elements are a non-homogenized combination of solid and void materials, as demonstrated in Figure 3.1(c). Representations of intermediate elements with a relative value of $X_e = 0.5$ in the two algorithms are shown in Figure 3.3. Areas 1 to 4 in an element with an elemental density of 0.5 (Figure 3.3a) have the identical material properties in SIMP; however, the same areas in an element with an elemental volume fraction of 0.5 (Figure 3.3b) have different material properties in SEMDOT. In Figure 3.3(b), the diagonal boundary across the element, which has a volume fraction of $X_e = 0.5$, is only a single instantiation of an element with this volume fraction. During optimization, the solid/void boundary associated with solid grid points can be imposed across each element that is on or near the solid/void boundary. Here, elemental stiffness matrices are estimated using a linear interpolation between the solid and void phases:

$$\begin{aligned} \mathbf{K}_e(X_e) &= (1 - X_e)\mathbf{K}_e^0 + X_e \mathbf{K}_e^1 \\ &= (1 - X_e)\rho_{\min}\mathbf{K}_e^1 + X_e \mathbf{K}_e^1, \end{aligned} \tag{3.5}$$

where $\mathbf{K}_e^0$ is the stiffness matrix of the void element.

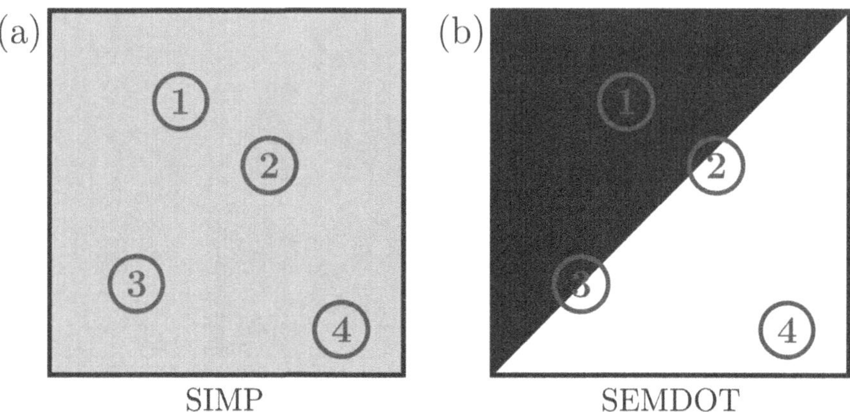

FIGURE 3.3
Representations of single intermediate elements in (a) SIMP and (b) SEMDOT with $X_e = 0.5$.

One of the problems in topology optimization is the checkerboard pattern problem. An effective way to solve this problem, and to ensure regularity or existence of topological design solutions, is to use filters to restrict the designs in the topology optimization algorithm [97]. The basic idea of a filter is to substitute a non-regular function with its regularized alternative [97]. The filter used for elemental volume fractions X_e is:

$$\tilde{X}_e = \frac{\sum_{l=1}^{N_e} \omega_{el} X_l}{\sum_{l=1}^{N_e} \omega_{el}}, \tag{3.6}$$

where $\tilde{X}_e$ is the filtered elemental volume fraction, N_e is the neighborhood set of elements within the filter domain for the eth element that is a circle centered at the centroid of this element with a predefined filter radius $r_{\min}$, and ω_{el} is a linear weight factor defined as:

$$\omega_{el} = \max(0, r_{\min} - \Delta(e, l)), \tag{3.7}$$

where $\Delta(e, l)$ is the center-to-center distance of the lth element within the filter domain to the eth element. The filtering technique presented in Equations (3.6) and (3.7) is illustrated in Figure 3.4, where the elements in the circular area are involved in the calculation of the filtered elemental volume fraction related to the centered element.

The key of forming smooth topological boundaries is to assign elemental volume fractions to grid points. As converting filtered elemental volume fractions to grid points directly requires high computational cost, filtered elemental volume fractions are transferred to nodes first with a heuristic filter, and then the obtained nodal densities are assigned to grid points using shape functions. Nodal densities are obtained by:

$$\rho_n = \frac{\sum_{e=1}^{M} \omega_{ne} \tilde{X}_e}{\sum_{e=1}^{M} \omega_{ne}}, \tag{3.8}$$

where ρ_n is the density of the nth node, M is the total number of elements, and ω_{ne} is the weight factor defined as:

$$\omega_{ne} = \max(0, 1 - \Delta(n, e)), \tag{3.9}$$

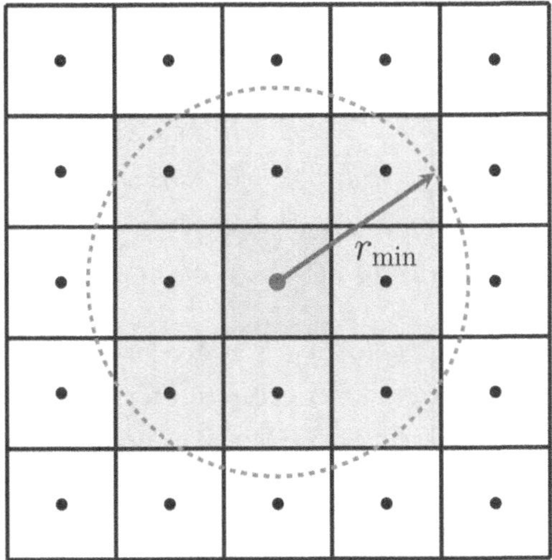

FIGURE 3.4
Illustration of filtering technique.

where $\Delta(n, e)$ is the distance between the nth node and the center of the eth element.

Considering a four-node element for 2D cases, densities of grid points $\rho(\zeta, \eta)$ are calculated using the linear interpolation of nodal densities ρ_n which has the form:

$$\rho(\zeta, \eta) = \sum_{\gamma=1}^{4} N^{\gamma}(\zeta, \eta)\rho_n^{\gamma}, \qquad \rho(\zeta, \eta) \in \rho(x, y), \tag{3.10}$$

where (ζ, η) is the local coordinate of the grid point, ρ_n^{γ} is the density for the γth node of the element, $N^{\gamma}(\zeta, \eta)$ is an appropriate shape function, (x, y) is the global coordinate of grid points, and $\rho(x, y)$ is the density of the grid point at (x, y).

For 3D cases, eight-node linear cubes are used, and the density of a grid point in an element is expressed with the shape function as:

$$\rho(\zeta, \eta, \xi) = \sum_{\gamma=1}^{8} N^{\gamma}(\zeta, \eta, \xi)\rho_n^{\gamma}, \qquad \rho(\zeta, \eta, \xi) \in \rho(x, y, z), \tag{3.11}$$

where (ζ, η, ξ) is the local coordinate of the grid point, $N^{\gamma}(\zeta, \eta, \xi)$ is the shape function, (x, y, z) is the global coordinate of grid points, and $\rho(x, y, z)$ is the density of the grid point at (x, y, z).

The Heaviside smooth function is used to obtain solid/void grid points, which is:

$$\rho_{e,g} = \begin{cases} \dfrac{\tanh(\beta\,\Psi) + \tanh[\beta\,(\rho(x, y) - \Psi)]}{\tanh(\beta\,\Psi) + \tanh[\beta\,(1.0 - \Psi)]} & \text{for 2D cases,} \\[2ex] \dfrac{\tanh(\beta\,\Psi) + \tanh[\beta\,(\rho(x, y, z) - \Psi)]}{\tanh(\beta\,\Psi) + \tanh[\beta\,(1.0 - \Psi)]} & \text{for 3D cases,} \end{cases} \tag{3.12}$$

where Ψ is a threshold value and β is a scaling parameter that controls the steepness and is updated by:

$$\beta_k = \beta_{k-1} + \Lambda, \tag{3.13}$$

where k is the current iteration number and Λ is the evolution rate for β.

Smooth topological boundaries of 2D cases can be implicitly extracted with the zero level-set $\Phi(x, y) = 0$ of the level-set function as:

$$\Phi(x,y) = \begin{cases} \rho(x,y) - \Psi > 0 & \text{for solid region,} \\ \rho(x,y) - \Psi = 0 & \text{for boundary,} \\ \rho(x,y) - \Psi < 0 & \text{for void region.} \end{cases} \tag{3.14}$$

Smooth topological boundaries of 3D cases are implicitly represented via a level-set function $\Phi(x, y, z)$:

$$\Phi(x,y,z) = \begin{cases} \rho(x,y,z) - \Psi > 0 & \text{for solid region,} \\ \rho(x,y,z) - \Psi = 0 & \text{for boundary,} \\ \rho(x,y,z) - \Psi < 0 & \text{for void region.} \end{cases} \tag{3.15}$$

The threshold value Ψ in the level-set function (Equations 3.14 and 3.15) is determined using the bi-section method such that the target volume constraint can be satisfied iteratively. It is noted that the value of Ψ is floating to meet the volume constraint during the optimization procedure.

For the next iteration of finite element analysis, filtered elemental volume fractions are updated by assembling grid points for their corresponding elements:

$$\tilde{X}_e^{new} = \frac{1}{N} \sum_{g=1}^{N} \rho_{e,g}^{new}, \tag{3.16}$$

where $\rho_{e,g}^{new}$ is the updated density of the grid point obtained by the Heaviside smooth function.

In SEMDOT, there are two convergence criteria used to terminate the optimization procedure: the overall topological alteration and topological boundary error. The overall topological alteration is defined as:

$$\frac{\sum_{e=1}^{M} \mid X_e^k - X_e^{k-1} \mid}{\sum_{e=1}^{M} X_e^k} \leq \tau, \tag{3.17}$$

where τ is the tolerance value for the overall topological alteration, which the authors' suggest to set at 0.001.

The topological boundary error convergence criterion is defined as:

$$\varepsilon = \frac{N_v}{M} \leq \epsilon, \tag{3.18}$$

where ε is the topological boundary error, N_v is the number of intermediate elements that are not along boundaries, and ϵ is the tolerance value for the topological boundary error, which the authors' suggest to set at 0.001.

The defined topological boundary error, ε, is schematically illustrated in Figure 3.5, where a design domain with 25 elements is taken as an example. Equation (3.18) is an indicator used to assess the accuracy of the smooth topological boundary formed by the level-set function (Equations 3.14 and 3.15). Generally, increasing the number of elements within the design domain or grid points within each element is a simple way to improve the boundary accuracy; however, increasing the number of elements or grid points per element will increase the computational cost.

The formation of accurate smooth boundaries in SEMDOT requires a process of reducing the boundary deviation iteratively, which is assessed by the topological boundary error ε

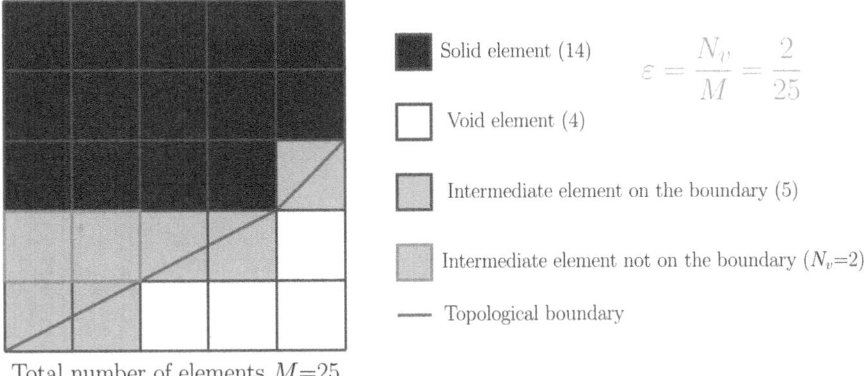

FIGURE 3.5
Illustration of topological boundary error.

(Equation 3.18). A high-boundary deviation related to Equation (3.18) or discontinuous features could be caused if the smooth boundary strategy in SEMDOT is directly used as the post-processing method for SIMP. That is, in the authors' opinion, the smooth-edged material distribution strategy proposed in SEMDOT should not be applied to the final topology obtained by SIMP as the post-processing method. The SEMDOT algorithm incrementally improves its smooth boundaries toward an optimum, and it is not a post-processing method for obtaining smooth boundaries from zigzag edge boundaries.

3.3 Sensitivity Analysis

Sensitivity analysis, that is, the derivative of the objective function with respect to the design variables, is required to determine the optimization direction when using gradient-based optimizers. Efficient and accurate sensitivity analysis is able to improve the convergence and efficiency of the algorithm [160]. Optimizers will be discussed in the next chapter, nevertheless they need the appropriate sensitivity information to appropriately update the topological design.

Assuming compliance as the objective function, which has the form of $C(X_e) = \mathbf{f}^{\mathbf{T}}\mathbf{u}$ where $\mathbf{f} = \mathbf{K}(X_e)\mathbf{u}$ and $\mathbf{K} = \sum_{e=1}^{M} \mathbf{K}_e(X_e)$ for solid and void elements, the sensitivity with respect to X_e can be directly calculated using the Discrete Variable Sensitivity strategy presented in [110–113]:

$$\frac{\partial C(X_e)}{\partial X_e} = \begin{cases} \dfrac{C(X_e = 1) - C(X_e = 0)}{1} \approx -\mathbf{u}_e^{\mathbf{T}}\mathbf{K}_e^1\mathbf{u}_e & \text{if } X_e = 1, \\ \dfrac{C(X_e = 0) - C(X_e = 1)}{-1} \approx -\rho_{\min}\mathbf{u}_e^{\mathbf{T}}\mathbf{K}_e^1\mathbf{u}_e & \text{if } X_e = \rho_{\min}, \end{cases} \tag{3.19}$$

where C is the objective function and $\mathbf{u}_e$ is the displacement vector of the eth element.

In SEMDOT, an intermediate element is defined as a boundary element that contains two materials, which can be generated by assembling solid ($\rho_{e,g} = 1$) and void ($\rho_{e,g} = \rho_{\min}$) grid points (see Figure 3.1c). The topological boundaries at the grid point level mean that there are no intermediate density grid points ($0 < \rho_{e,g} < 1$) because grid-points are either inside or outside the defined boundary. Here, physically meaningful sensitivities of the boundary

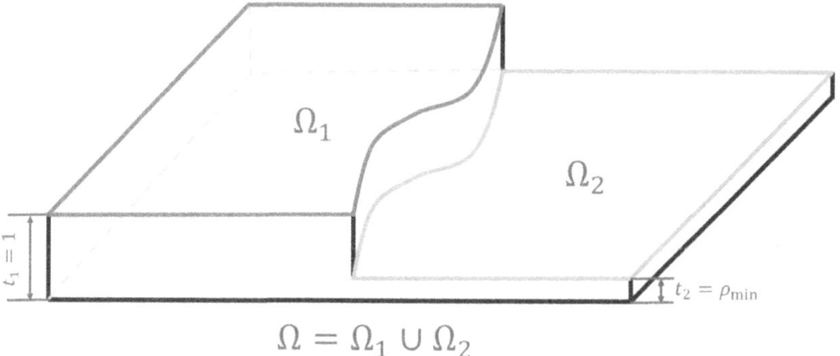

FIGURE 3.6
Schematic illustration of sensitivities of boundary elements.

elements are calculated using a linear combination of the sensitivities of void elements with a weighting factor of $(1 - X_e)$ and sensitivities of solid elements with a weighting factor of X_e, which has the form:

$$\begin{aligned}\frac{\partial C(X_e)}{\partial X_e} &\approx (1 - X_e)\frac{\partial C(X_e)}{\partial X_e}\bigg|_{X_e=\rho_{\min}} + X_e\frac{\partial C(X_e)}{\partial X_e}\bigg|_{X_e=1} \\ &= -[(1 - X_e)\rho_{\min} + X_e]\mathbf{u}_e^{\mathbf{T}}\mathbf{K}_e^1\mathbf{u}_e.\end{aligned} \tag{3.20}$$

Thanks to the Discrete Variable Sensitivity strategy [110–113], the material penalization scheme used in SIMP and BESO is not required in SEMDOT, and hence non-penalization topology optimization is realized. Most importantly, Equation (3.20) can be schematically illustrated by a plate with two different thicknesses ($t_1 = 1$ and $t_2 = \rho_{\min}$) shown in Figure 3.6, which is inspired by Sun et al. [168]. In Figure 3.6, the whole plate (Ω) is the combination of a region with the thickness of $t_1 = 1$ (Ω_1) and another region with the thickness of $t_2 = \rho_{\min}$ (Ω_2). Based on Figure 3.6, the sensitivities of boundary elements in SEMDOT can be estimated by:

$$\begin{aligned}\frac{\partial C(X_e)}{\partial X_e} &\approx \left(\frac{V_{\Omega_1} + V_{\Omega_2}}{V_t}\right)\frac{\partial C(X_e)}{\partial X_e}\bigg|_{X_e=1} \\ &= \left(\frac{A_{\Omega_1}t_1 + A_{\Omega_2}t_2}{A_\Omega t_1}\right)\frac{\partial C(X_e)}{\partial X_e}\bigg|_{X_e=1} \\ &= -[(1 - X_e)\rho_{\min} + X_e]\mathbf{u}_e^{\mathbf{T}}\mathbf{K}_e^1\mathbf{u}_e,\end{aligned} \tag{3.21}$$

where V_{Ω_1} and V_{Ω_2} are the volumes of domains Ω_1 and Ω_2, respectively; and V_t is the volume of the plate where the thickness is $t_1 = 1$; and A_{Ω_1}, A_{Ω_2}, and A_Ω are the areas of domains Ω_1, Ω_2, and Ω, respectively. In Equation (3.21), $\partial C(X_e)/\partial X_e|_{X_e=1}$ represents the sensitivity of the plate where the thickness is $t_1 = 1$. Figure 3.6 can also be used to demonstrate the Young's moduli and stiffness matrices of the two regions in SEMDOT.

When the filtering technique (Equation 3.6) is applied on X_e, sensitivities of the objective functional with respect to X_e can be calculated based on the chain rule as:

$$\frac{\partial C(\tilde{X}_e)}{\partial X_l} = \sum_{e=1}^{N_l}\frac{\partial C(\tilde{X}_e)}{\partial \tilde{X}_e}\frac{\partial \tilde{X}_e}{\partial X_l} = \sum_{e=1}^{N_l}\frac{\omega_{el}}{\sum\limits_{\varsigma=1}^{N_e}\omega_{e\varsigma}}\frac{\partial C(\tilde{X}_e)}{\partial \tilde{X}_e}. \tag{3.22}$$

It should be noted that material properties and sensitivity analysis of the boundary elements are not limited to linear combinations. Some non-linear combinations or other strategies (for example, high order sensitivity analysis) can also be considered to pursue more accurate sensitivities. This chapter only presents the simplest form to reduce the algorithm complexity and to make it easy for potential users to understand the algorithm mechanism.

The authors note that this same form, $\square = (1 - X_e)\square_{\text{Void}} + X_e\square_{\text{Solid}}$, for the assumption of material properties and sensitivities of boundary elements causes a less rigorous mathematical expression. Mathematically, sensitivities are calculated based on the derivative of the function of the stiffness matrix with respect to the design variable. Compared to traditional element-based algorithms, the SEMDOT approach provides a new way to explore

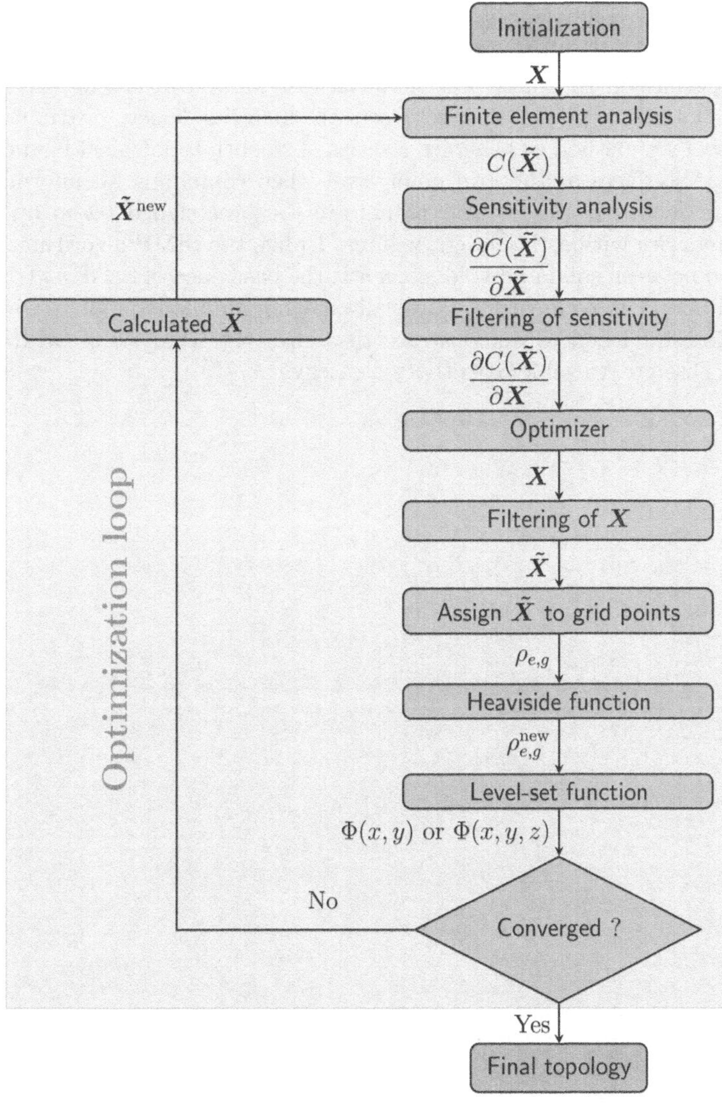

FIGURE 3.7
Flowchart of the SEMDOT method.

different topological designs. The effectiveness of the assumption of sensitivity analysis in SEMDOT will be validated in the following chapters.

For ease of understanding, the optimization procedure of the SEMDOT method is illustrated in Figure 3.7, where $\boldsymbol{X}$ and $\tilde{\boldsymbol{X}}$ are the vectors of X_e and $\tilde{X}_e$, respectively.

3.4 Summary

This chapter clearly demonstrated the motivation for proposing a new topology optimization algorithm that is capable of directly generating smooth boundaries and the importance of the non-penalization approach. The algorithm framework of the non-penalization SEMDOT algorithm is systematically demonstrated. The framework shows that SEMDOT runs at two levels: elements and grid points. At the element level, finite element analysis and sensitivity analysis are performed to update the material distribution in the design volume. At the grid-point level, the grid point values represent the material density distribution within each element. A level-set method to generate a series of smooth topological boundaries based on the solid/void interfaces at the grid-point level when combining all information from the design volume. The densities at grid points are then reassigned based on the solid/void interface boundaries within the design volume. Unlike the SIMP algorithm, which pursues the separation between solid and void elements, the basic idea of SEMDOT is to pursue the minimization boundary elements. A physically meaningful model, a plate with two different thicknesses, is established to demonstrate the sensitivity analysis of SEMDOT, which is based on the Discrete Variable Sensitivity strategy.

4

Optimizers

This chapter briefly introduces two of the most representative optimizers or optimization solvers: Optimality Criteria (OC) and Method of Moving Asymptotes (MMA). In general, optimizers are a crucial part of topology optimization, and they are employed to provide efficient and accurate updates to the topological design, and enable the shortest path to the optimal design. The OC and MMA optimizers will be used to solve the topology optimization problems in the following chapters. It should be noted that OC and MMA are all gradient-based algorithms. In some real-world cases (for example, crashworthiness design), it is difficult to calculate the sensitivities for the governing objective functions, and hence non-gradient optimization solvers (for example, Genetic Algorithm (GA)) are useful. However, this chapter will not take into account non-gradient optimization solvers since the sensitivity information can be easily obtained for the case studies that are considered.

4.1 Optimality Criteria (OC) Method

The Optimality Criteria (OC) method is a classical approach used to solve structural optimization problems. Despite being an old approach, the OC method is still widely used because of its numerical simplicity and efficiency. The OC method is formulated on the condition that if the constraint $0 \leq \boldsymbol{X} \leq 1$ is inactive, the convergence is reached following the Karush-Kuhn-Tucker (KKT) condition [118]:

$$\frac{\partial C(\boldsymbol{X})}{\partial X_e} + \lambda \frac{\partial \bar{v}(\boldsymbol{X})}{\partial X_e} = 0, \tag{4.1}$$

where C is the objective function, X_e is the design variable, $\boldsymbol{X}$ is the vector of X_e, and λ is the Lagrange multiplier associated with the constraint $\bar{v}(\boldsymbol{X})$.

This optimality condition can be expressed as $\mathbb{B}_e = 1$, which is:

$$\mathbb{B}_e = -\frac{\partial C(\boldsymbol{X})}{\partial X_e} \left(\lambda \frac{\partial \bar{v}(\boldsymbol{X})}{\partial X_e} \right)^{-1}. \tag{4.2}$$

It is noted that the Lagrange multiplier λ in Equation (4.2) can be determined by a simple bi-sectioning strategy. The bi-sectioning strategy can find where a function is equal to zero if two known function values are of opposite signs. The assumption is the function must pass through zero between the two known function values. The search segments the search space into two, calculates the mid-point function value, and selects the segment that continues to have two known function values that changes sign. This search continues until the segment size is smaller than a tolerance level or a segment function value equals zero or maximum iterations are reached. The linear Taylor series expansion can be adopted to

DOI: 10.1201/9781032634449-4

determine if the equality constraint is satisfied, which is expressed by:

$$C^k + \sum_{e=1}^{M} \left.\frac{\partial C}{\partial X_e}\right|_{X_e^k} [X^{k+1}(\lambda) - X^k] \approx 0, \tag{4.3}$$

or

$$C^k + \sum_{e=1}^{M} \left.\frac{\partial C}{\partial X_e}\right|_{X_e^k} \frac{X^k}{X^{k+1}(\lambda)} [X^{k+1}(\lambda) - X^k] \approx 0. \tag{4.4}$$

From the numerical computation point of view, Equation (4.4) is generally utilized to determine if the equality constraint is met because it is a better approximation [126].

The scheme used to update design variables, proposed by Bendsøe [18], is expressed by:

$$X_e^{new} = \begin{cases} \max(0, X_e - mov) & \text{if } X_e \mathbb{B}_e^{Dam} \leq \max(0, X_e - mov), \\ \min(1, X_e + mov) & \text{if } X_e \mathbb{B}_e^{Dam} \geq \min(1, X_e + mov), \\ X_e \mathbb{B}_e^{Dam} & \text{Otherwise}, \end{cases} \tag{4.5}$$

where X_e^{new} is the new design variable, mov is a positive definite valued move limit, and Dam is a numerical damping coefficient. In OC, mov is the main parameter that can be adjusted to yield various topological designs. The extension of standard OC to obtain black-and-white designs can be found in [64]. As the goal of the Smooth-Edged Material Distribution for Optimizing Topology (SEMDOT) algorithm is not to achieve pure black-and-white designs, this extended OC will not be discussed here. However, the extended OC paper is a good reference material for the readers.

4.2 Method of Moving Asymptotes (MMA)

The Method of Moving Asymptotes (MMA) proposed by Svanberg [169], a variant of CONLIN [44], is a mathematical programming algorithm that is extensively used for topology optimization. MMA works with a sequence of approximate sub-problems of given type. In MMA, the aforementioned sub-problems are separable and convex. Sensitivity information at the current iteration and some of the iteration history are required to construct the sub-problems. The sub-problem is solved at each iteration using a dual method or an interior point algorithm (primal-dual algorithm), and then the solution to the sub-problem is used for the next design. The dual method makes use of a continuous convex barrier function that approximates the objective function plus the constraint function to provide a system where the gradient will enable a path to the approximate optimal solution.

Similar with CONLIN, the MMA approximation is a first order approximation. The MMA approximation of an objective functional C with M design variables (elemental volume fractions $\boldsymbol{X} = (X_1, \cdots, X_M)$) subjected to a given iteration point $\boldsymbol{X}^0$ is expressed by:

$$C(\boldsymbol{X}) \approx C(\boldsymbol{X}^0) + \sum_{e=1}^{M} \left(\frac{\bar{r}_e}{\bar{U}_e - X_e} + \frac{\bar{s}_e}{X_e - \bar{L}_e} \right), \tag{4.6}$$

where $\bar{L}_e$ and $\bar{U}_e$ are lower and upper moving asymptotes that are updated iteratively to mitigate oscillation or improve convergence, and the parameters $\bar{r}_e$ and $\bar{s}_e$ are defined as:

$$\begin{aligned} &\text{if } \frac{\partial C}{\partial X_e}(X^0) > 0 \text{ then } \bar{r}_e = (\bar{U}_e - X_e^0)^2 \frac{\partial C}{\partial X_e}(X^0) \text{ and } \bar{s}_e = 0, \\ &\text{if } \frac{\partial C}{\partial X_e}(X^0) < 0 \text{ then } \bar{r}_e = 0 \text{ and } \bar{s}_e = -(X_e^0 - \bar{L}_e)^2 \frac{\partial C}{\partial X_e}(X^0). \end{aligned} \tag{4.7}$$

A heuristic scheme proposed by Svanberg [169] is used to update the asymptotes. In this scheme, if a design variable oscillates, $\bar{L}_e$ and $\bar{U}_e$ are moved closer to X_e. This can increase the convexity of MMA approximations and stabilize the convergence process. If the convergence rate is low, $\bar{L}_e$ and $\bar{U}_e$ are moved further away from X_e. This can decrease the convexity of MMA approximations and accelerate the convergence process. In the first two iterations, the scheme is given by:

$$\begin{aligned} \bar{L}_e^k &= X_e^k - S_{ini}(X_{ue} - X_{le}), \\ \bar{U}_e^k &= X_e^k + S_{ini}(X_{ue} - X_{le}), \end{aligned} \tag{4.8}$$

where k is the current iteration, S_{ini} is the initial parameter, and X_{ue} and X_{le} are the upper and lower bounds of each design variable, respectively.

From the third iteration, if $(X_e^k - X_e^{k-1})(X_e^{k-1} - X_e^{k-2}) < 0$, we have:

$$\begin{aligned} \bar{L}_e^k &= X_e^k - S_{decr}(X_e^{k-1} - \bar{L}_e^{k-1}), \\ \bar{U}_e^k &= X_e^k + S_{decr}(\bar{U}_e^{k-1} - X_e^{k-1}). \end{aligned} \tag{4.9}$$

If $(X_e^k - X_e^{k-1})(X_e^{k-1} - X_e^{k-2}) > 0$, we have:

$$\begin{aligned} \bar{L}_e^k &= X_e^k - S_{incr}(X_e^{k-1} - \bar{L}_e^{k-1}), \\ \bar{U}_e^k &= X_e^k + S_{incr}(\bar{U}_e^{k-1} - X_e^{k-1}), \end{aligned} \tag{4.10}$$

where S_{decr} is the tightening factor and S_{incr} is the relaxation factor.

To solve the sub-problem efficiently, the Lagrange functional of the approximated problem in the dual method is given by:

$$\mathcal{L}^k(\boldsymbol{X}, \boldsymbol{\lambda}) = C_0^k(\boldsymbol{X}) + \boldsymbol{\lambda}^{\mathrm{T}} \boldsymbol{g}^k(\boldsymbol{X}), \tag{4.11}$$

where $\mathcal{L}^k(\boldsymbol{X}, \boldsymbol{\lambda})$ is the Lagrangian of the sub-problem, $C_0^k(\boldsymbol{X})$ represents the convex and separable approximations of the primal objective function, $\boldsymbol{\lambda}$ represents the corresponding Lagrange multipliers, and $\boldsymbol{g}(\boldsymbol{X})$ represents the constraints.

The efficiency of the dual method highly depends on the convexity and separability of the approximations. This ensures that $\mathcal{L}^k(\boldsymbol{X}, \boldsymbol{\lambda})$ is also convex and separable. The dual of the Lagrangian is expressed by:

$$C_d^k(\boldsymbol{\lambda}) = \min_{\boldsymbol{X} \in \mathcal{B}} \mathcal{L}^k(\boldsymbol{X}, \boldsymbol{\lambda}) = \min_{\boldsymbol{X} \in \mathcal{B}} \{C_0^k(\boldsymbol{X}) + \boldsymbol{\lambda}^{\mathrm{T}} \boldsymbol{g}^k(\boldsymbol{X})\}, \tag{4.12}$$

where $\mathcal{B}$ is the set of design variables.

The optimal solutions can be obtained by:

$$\boldsymbol{X}^{k+1} = \arg \min_{\boldsymbol{X} \in \mathcal{B}} \mathcal{L}^k(\boldsymbol{X}, \boldsymbol{\lambda}). \tag{4.13}$$

The optimal solutions $\boldsymbol{X}^{k+1}$ can be obtained by solving the first-order stationary conditions:

$$\nabla C_0^k(\boldsymbol{X}) + \boldsymbol{\lambda}^{\mathrm{T}} \nabla \boldsymbol{g}^k(\boldsymbol{X}) = 0. \tag{4.14}$$

Using $\boldsymbol{X}^{k+1}$ in Equation(4.12), the dual function can be therefore given by:

$$C_d^k(\boldsymbol{\lambda}) = \mathcal{L}^k(\boldsymbol{X}^{k+1}(\boldsymbol{\lambda}), \boldsymbol{\lambda}) = C_0^k(\boldsymbol{X}^{k+1}(\boldsymbol{\lambda})) + \boldsymbol{\lambda}^{\mathrm{T}} \boldsymbol{g}^k(\boldsymbol{X}^{k+1}(\boldsymbol{\lambda})). \tag{4.15}$$

For simple optimization problems, MMA may be slower than the OC method, but for more sophisticated optimization problems with several constraints, MMA shows stronger convergence properties [20]. Other than standard MMA, there are some improved versions, for example, Globally Convergent MMA (GCMMA) [170] and Two-point Gradient-based MMA (TGMMA) [105]. Although there are some more advanced mathematical programming tools such as SNOPT [61] and IPOPT [177], there is no solid evidence that they are more efficient or more reliable than MMA. Therefore, MMA has been widely accepted as a standard optimizer for structural topology optimization [163]. The MATLAB codes of MMA are available (mmasub and subsolv). The readers can obtain the codes by contacting Prof. Krister Svanberg (krille@math.kth.se) from KTH in Stockholm, Sweden. Although MMA has much more input and output variables in comparison with OC, the strategy of properly choosing MMA parameters is well-established in the topology optimization community. It should be mentioned that the BESO method cannot use MMA, and the OC-like approach is being used in BESO as the default optimizer. Generally, MMA is used as the default optimizer for SIMP, MMC-based, and SEMDOT algorithms.

The level-set method introduced in Chapter 2 also uses regularization and approximations to ensure consistency and convexity in the update of the level-set objective function, and therefore, the topological design update. It should be noted that OC and MMA optimizers are not used in the level-set method introduced in Chapter 2.

Topology optimization solvers are not limited to OC and MMA. There are some other well-established optimizers, including Sequential Linear Programming (SLP), Sequential Quadratic Programming (SQP), and projected subgradient method (an algorithm similar to the steepest descent method). More details on optimization solvers can be found in [123]. This book only concentrates on the use of OC and MMA in the SEMDOT algorithm. Different topology optimization algorithms provide various strategies to discover topological designs, and the optimization solvers are the engine that is used to drive optimization. The parameters in optimization solvers have significant effect on final results.

4.3 Summary

This chapter introduced two typical optimization solvers, OC and MMA, that are extensively used in topology optimization for different applications. The algorithm mechanisms of these two optimizers are demonstrated in detail. The MMA optimizer is generally recommended because of its superior performance in solving more sophisticated optimization problems with several constraints.

5

Stiffness Maximization Problems

This chapter solves the classic stiffness maximization problems using Smooth-Edged Material Distribution for Optimizing Topology (SEMDOT). In this chapter, the influences of the move limit variable in the Optimality Criteria (OC) and Method of Moving Asymptotes (MMA) optimizers, the number of grid points in each element, and filter radius on the final results are comprehensively investigated via both 2D and 3D cases. In addition, multiple load cases and optimization problems with passive void and solid elements are discussed. Two real-world cases (that is, Sheikh-Ibrahim's steel beam splice connection and suspension support rod of a vehicle's rear axle) are also presented in this chapter.

5.1 Optimization Problem

The compliance minimization or stiffness maximization case, the most widely tested benchmark problem in topology optimization, is considered here to demonstrate the effectiveness of Smooth-Edged Material Distribution for Optimizing Topology (SEMDOT). The optimization problem is stated as:

$$
\begin{aligned}
\min : & \; C(X_e) = \mathbf{f}^{\mathbf{T}}\mathbf{u} \\
\text{subject to} : & \; \mathbf{K}(X_e)\mathbf{u} = \mathbf{f}, \\
& \frac{\sum\limits_{e=1}^{M} X_e V_e}{\sum\limits_{e=1}^{M} V_e} - V^* \leq 0, \\
& 0 < \rho_{\min} \leq X_e \leq 1; \; e = 1, 2, \cdots, M,
\end{aligned}
\tag{5.1}
$$

where $C(X_e)$ is the objective function; X_e is the elemental volume fraction; $\mathbf{f}$ and $\mathbf{u}$ are global force and displacement vectors, respectively; $\mathbf{K}(X_e)$ is the global stiffness matrix; M is the total number of elements; V_e is the volume of the eth element; V^* is the target volume; and $\rho_{\min}$ is a small value, which is 0.001 here.

Sensitivities of the solid and void elements are calculated by:

$$
\frac{\partial C(X_e)}{\partial X_e} = \begin{cases} \dfrac{C(X_e = 1) - C(X_e = 0)}{1} \approx -\mathbf{u}_e^{\mathbf{T}}\mathbf{K}_e^1\mathbf{u}_e & \text{if } X_e = 1, \\ \dfrac{C(X_e = 0) - C(X_e = 1)}{-1} \approx -\rho_{\min}\mathbf{u}_e^{\mathbf{T}}\mathbf{K}_e^1\mathbf{u}_e & \text{if } X_e = \rho_{\min}, \end{cases}
\tag{5.2}
$$

where $\mathbf{u}_e$ is the displacement vector of the eth element and $\mathbf{K}_e^1$ is the stiffness matrix of the solid element.

DOI: 10.1201/9781032634449-5

Sensitivities of boundary elements are estimated as:

$$\begin{aligned}\frac{\partial C(X_e)}{\partial X_e} &\approx (1-X_e)\left.\frac{\partial C(X_e)}{\partial X_e}\right|_{X_e=\rho_{\min}} + X_e\left.\frac{\partial C(X_e)}{\partial X_e}\right|_{X_e=1} \\ &= -[(1-X_e)\rho_{\min} + X_e]\mathbf{u}_e^{\mathbf{T}}\mathbf{K}_e^1\mathbf{u}_e.\end{aligned} \tag{5.3}$$

5.2 Numerical Examples

5.2.1 Effects of Varying the Move Limit in OC and Smooth-Edged Material Distribution for Optimizing Topology (SEMDOT)

This section will investigate the effects of varying the move limit mov in OC and Method of Moving Asymptotes (MMA) on the structural performance and resulting topologies, this is because the move limit is one of the key parameters for optimization solvers. Both 2D and 3D short cantilever beam cases are considered as examples as shown in Figures 5.1 and 5.2, respectively. In Figure 5.1, the left side is fully constrained and a unit vertical load ($F = -1$ N) is imposed at the center point of the right side. In the 2D case, a mesh of 100×50 is used, the target volume V^* is set to 0.5, and the filter radius $r_{\min}$ is set to 1.5. In Figure 5.2, the design domain is fixed in one end, and a unit distributed vertical load is applied downwards on the lower free edge. In the 3D case, a mesh of 60×30×20 is used, V^* is set to 0.3, and $r_{\min}$ is set to 2.

The convergence process and optimized topology for the 2D short cantilever beam when $mov = 0.2$ in OC are shown in Figure 5.3. Figure 5.3(a) shows that the optimization process converges at a compliance value of 61.4541 after 116 iterations, and the topological boundary error reaches a very small value close to 0. The optimized topology is shown in Figure 5.3(b).

Convergence processes of the optimization of the 2D short cantilever beam under a range of move limits ($mov = 0.1$, $mov = 0.15$, $mov = 0.25$, $mov = 0.3$, $mov = 0.35$, and $mov = 0.4$) in OC are shown in Figure 5.4. It can be seen that the effect of the move limit in OC on the structural performance is insignificant. By contrast, the total number of iterations before convergence does vary with the move limit. However, there is no clear relationship between the move limit and total number of iterations before convergence. When changing the move limit, it can be seen that different topologies can arise, as seen in Figure 5.5. The topology designs for Figures 5.5(c) to (f) are very similar with a large void to the right of the beam, and triangular voids above and below the centerline of the beam, and very small void in the center left of the beam. The topology designs for Figures 5.5(a) and (b) have a small three void cluster to the left of center of the beam.

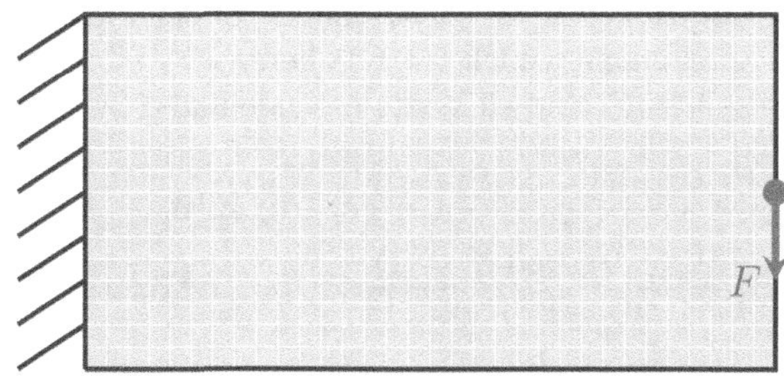

FIGURE 5.1
2D short cantilever beam.

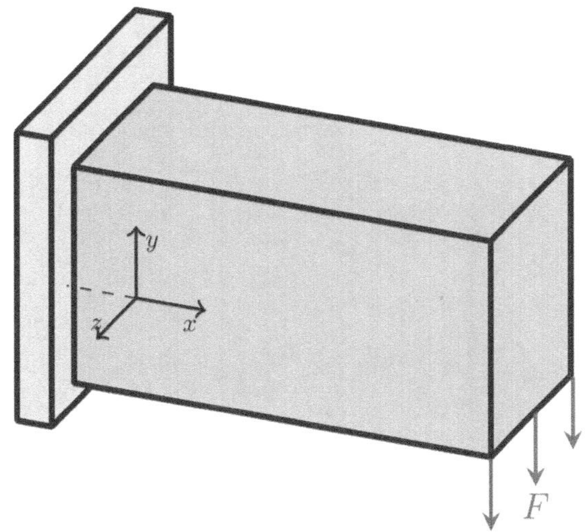

FIGURE 5.2
3D short cantilever beam.

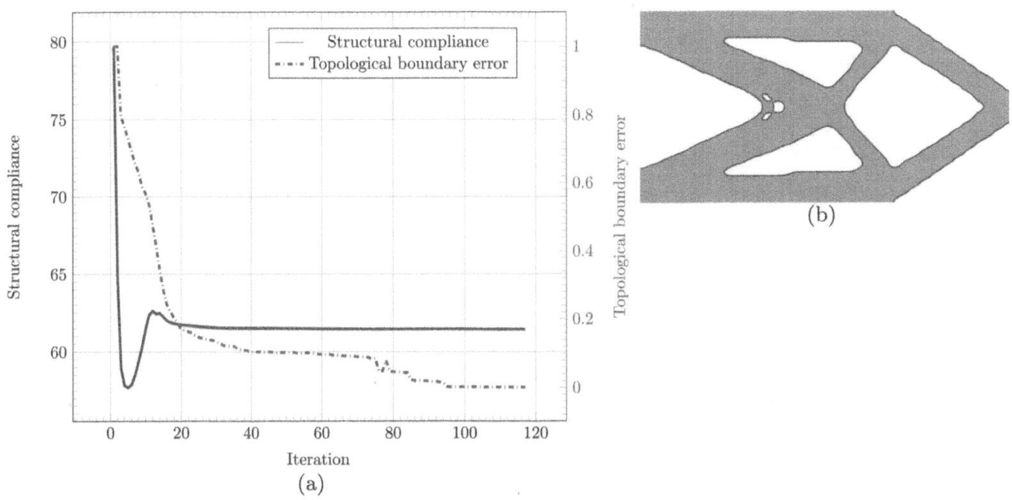

FIGURE 5.3
2D short cantilever beam: (a) convergence process and (b) optimized topology when *mov* = 0.2 in OC.

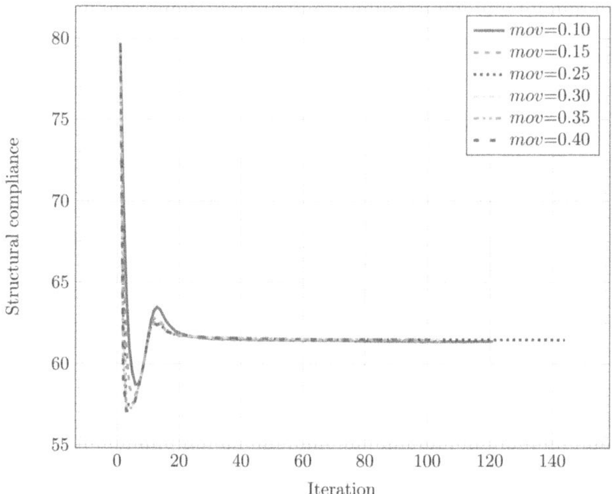

FIGURE 5.4
Comparison of convergence processes of 2D short cantilever beam under different move limits in OC.

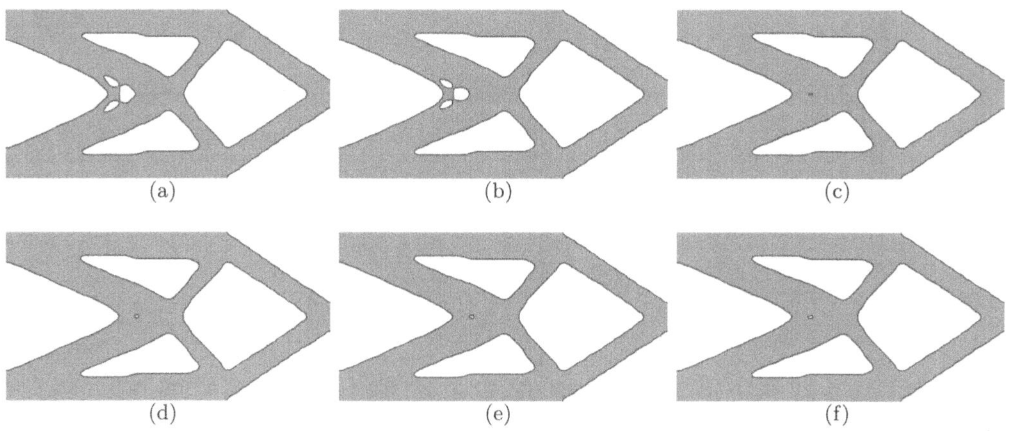

FIGURE 5.5
Resulting topologies of 2D short cantilever beam under different move limits in OC: (a) mov = 0.1, (b) mov = 0.15, (c) mov = 0.25, (d) mov = 0.3, (e) mov = 0.35, and (f) mov = 0.4.

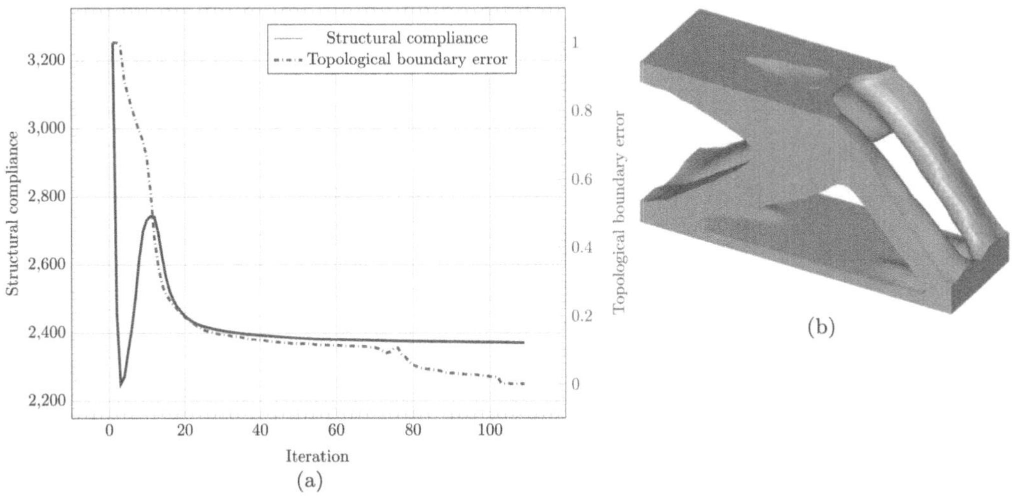

FIGURE 5.6
3D short cantilever beam: (a) convergence process and (b) optimized topology when *mov* = 0.2 in OC.

The convergence process and optimized topology for the 3D short cantilever beam when *mov* = 0.2 in OC are shown in Figure 5.6. Figure 5.6(a) shows that the optimization process converges at a compliance value of 2,370.4079 after 109 iterations, and the topological boundary error reaches 0. The optimized topology is shown in Figure 5.6(b).

Convergence processes of the optimization of the 3D short cantilever beam under a range of move limits (*mov* = 0.1, *mov* = 0.15, *mov* = 0.25, *mov* = 0.3, *mov* = 0.35, and *mov* = 0.4) in OC are shown in Figure 5.7. In this case, both the structural compliance and the total number of iterations before convergence are not significantly affected by varying the move limit. As shown in Figure 5.8, the obtained topological designs under different move limits are almost the same.

When the MMA optimizer is used to update the topological configuration iteratively for the compliance minimization cases, a dual method is adopted to solve the optimization problem. The optimization problem is then converted to solve the Lagrange function:

$$\mathcal{L} = C(X_e^k) - \sum_{e=1}^{M} \frac{(X_e^k - L_e)^2}{X_e - L_e} \frac{\partial C(X_e^k)}{\partial X_e} + \lambda \left(\sum_{e=1}^{M} V_e X_e - V^* \right), \tag{5.4}$$

where k is the current iteration number, X_e^k is the design variable at the kth iteration, and L_e is the lower moving asymptote. In Equation (5.4), the value of λ is adjusted to update the design variable X_e^{k+1} such that the volume constraint can be satisfied. When changing the value of the move limit, the other parameters in MMA are kept unchanged. Generally, the parameters are: epsimin=10^{-7}, raa0 = 0.00001, albefa = 0.1, asyinit = 0.1, asyincr = 1.2, and asydecr = 0.7. Detailed explanations about these parameters can be found in [171].

The convergence process and optimized topology for the 2D short cantilever beam when *mov* = 0.5 in MMA are shown in Figure 5.9. Figure 5.9(a) shows that the optimization process converges at a compliance value of 60.5888 after 118 iterations, and the topological boundary error reaches a negligible value. The resulting topology is shown in Figure 5.9(b).

Convergence processes of the optimization of the 2D short cantilever beam under a range of move limits (*mov* = 0.1, *mov* = 0.35, *mov* = 0.75, *mov* = 1, *mov* = 1.5, and *mov* = 2) in MMA are shown in Figure 5.10. It can be seen that varying the move limit has almost no

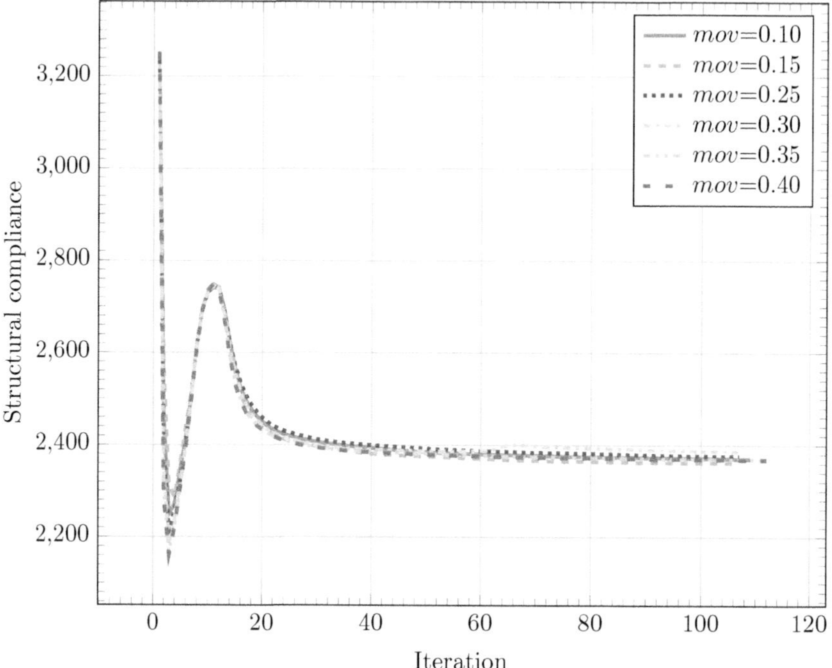

FIGURE 5.7
Comparison of convergence processes of 3D short cantilever beam under different move limits in OC.

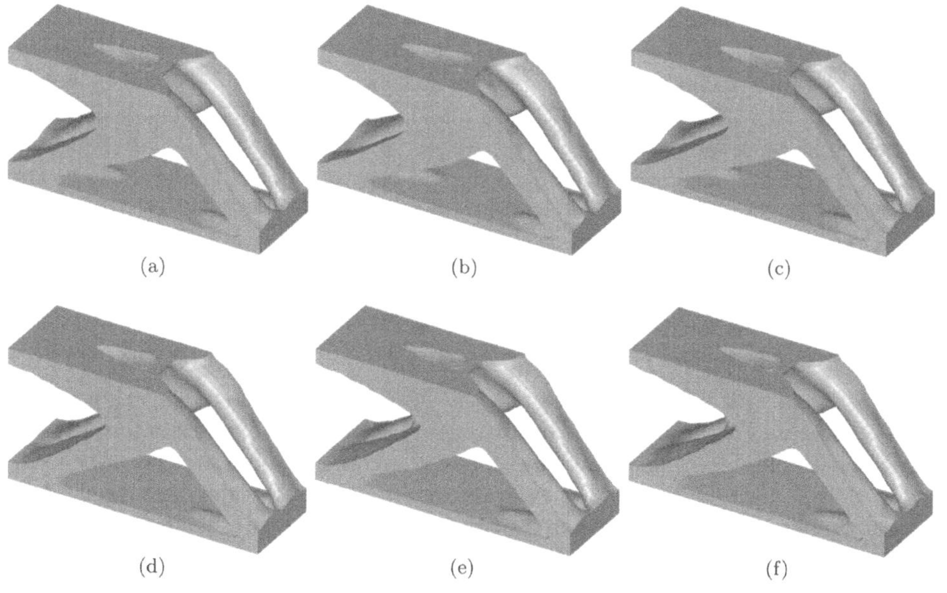

FIGURE 5.8
Resulting topologies of 3D short cantilever beam under different move limits in OC: (a) $mov = 0.1$, (b) $mov = 0.15$, (c) $mov = 0.25$, (d) $mov = 0.3$, (e) $mov = 0.35$, and (f) $mov = 0.4$.

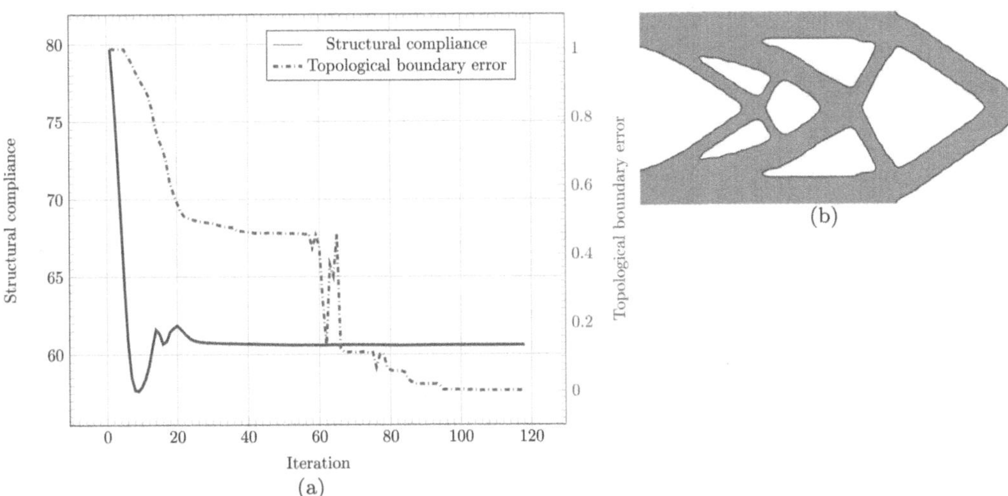

FIGURE 5.9
2D short cantilever beam: (a) convergence process and (b) optimized topology when *mov* = 0.5 in MMA.

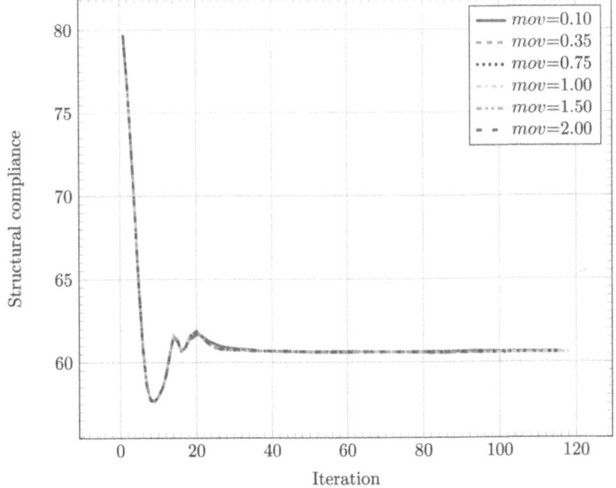

FIGURE 5.10
Comparison of convergence processes of 2D short cantilever beam under different move limits in MMA.

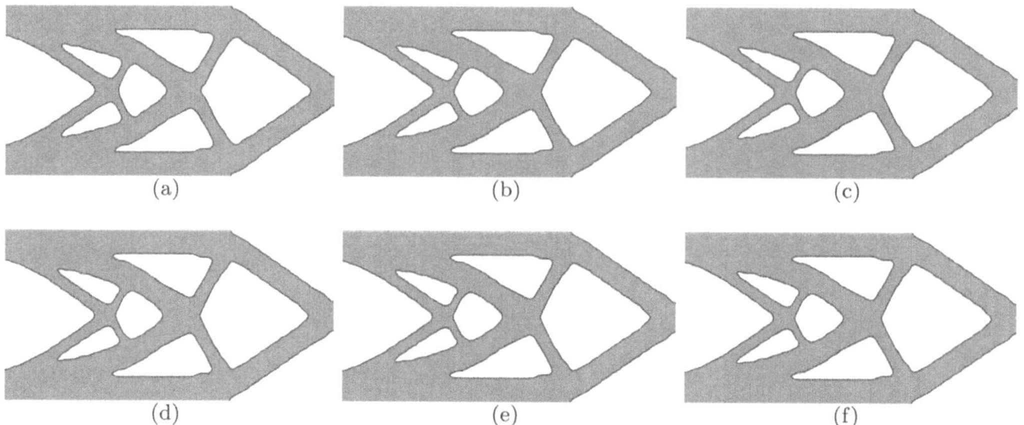

FIGURE 5.11
Resulting topologies of 2D short cantilever beam under different move limits in MMA: (a) $mov = 0.1$, (b) $mov = 0.35$, (c) $mov = 0.75$, (d) $mov = 1$, (e) $mov = 1.5$, and (f) $mov = 2$.

influence on both the structural performance and the total number of iterations. Similarly, Figure 5.11 shows that the obtained topologies under all the move limits are almost the same, where the large void is on the right of the beam, the two triangular voids are above and below the centerline, and there is a mid-sized three void cluster to the right of the center of the beam.

The convergence process and optimized topology for the 3D short cantilever beam when $mov = 0.5$ in MMA are shown in Figure 5.12. Figure 5.12(a) shows that the optimization process converges at a compliance value of 2,267.5876 after 104 iterations, and the topological boundary error reaches a negligible value. The resulting topology is shown in Figure 5.12(b).

Convergence processes of the optimization of the 3D short cantilever beam under a range of move limits ($mov = 0.1$, $mov = 0.35$, $mov = 0.75$, $mov = 1$, $mov = 1.5$, and $mov = 2$) in MMA are shown in Figure 5.13. When the move limit is set as $mov = 0.1$, the total number of iterations is 138, which is longer than the other cases. Otherwise, the move limit variable has only a very slight effect on the structural performance and the total number of iterations. The topological designs shown in Figure 5.14 are almost the same.

Based on the above discussion, it can be concluded that the SEMDOT algorithm is not sensitive to the move limit variable in OC and MMA. Compared to OC, MMA can provide improved solutions (lower structural compliance) and is generally less sensitive to the move limit for the 2D and 3D cantilever beam examples. Therefore, the authors have used MMA as the default optimizer for the SEMDOT algorithm. Unless otherwise stated, the MMA optimizer will be used for the rest of case studies, and the move limit of $mov = 1$ is adopted.

5.2.2 Effects of Grid Points

In contrast with representative element-based approaches such as SIMP and BESO, the SEMDOT algorithm has grid points associated with each element to realize the smooth topological design. This section will study the influences of the number of grid points per element on the structural performance, topological configuration, convergence and computational efficiency. Both 2D and 3D long cantilever beam cases are considered as shown in

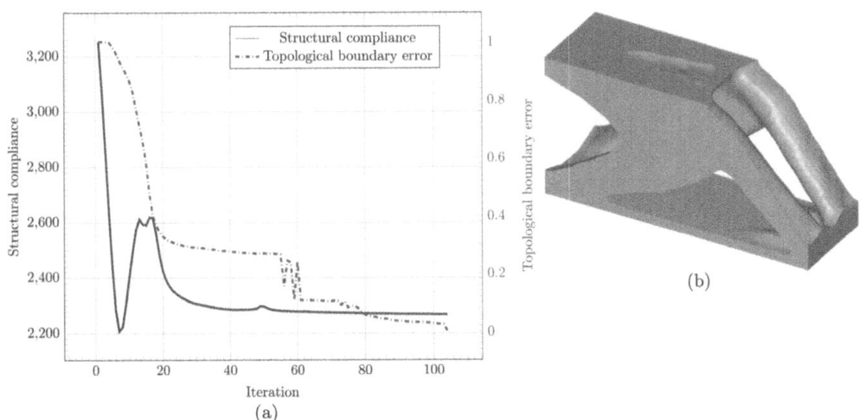

FIGURE 5.12
3D short cantilever beam: (a) convergence process and (b) optimized topology when *mov* = 0.5 in MMA.

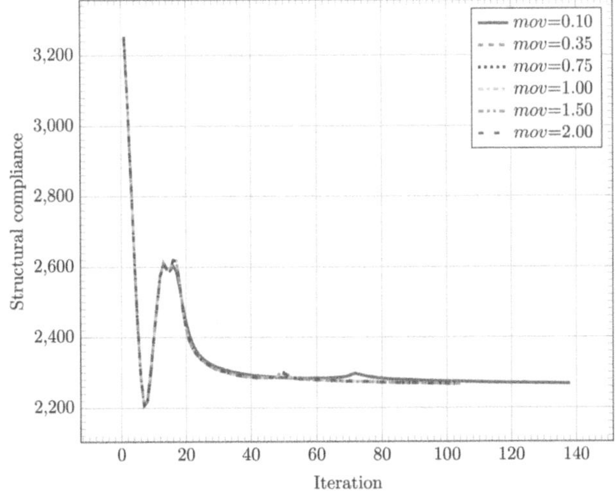

FIGURE 5.13
Comparison of convergence processes of 3D short cantilever beam under different move limits in MMA.

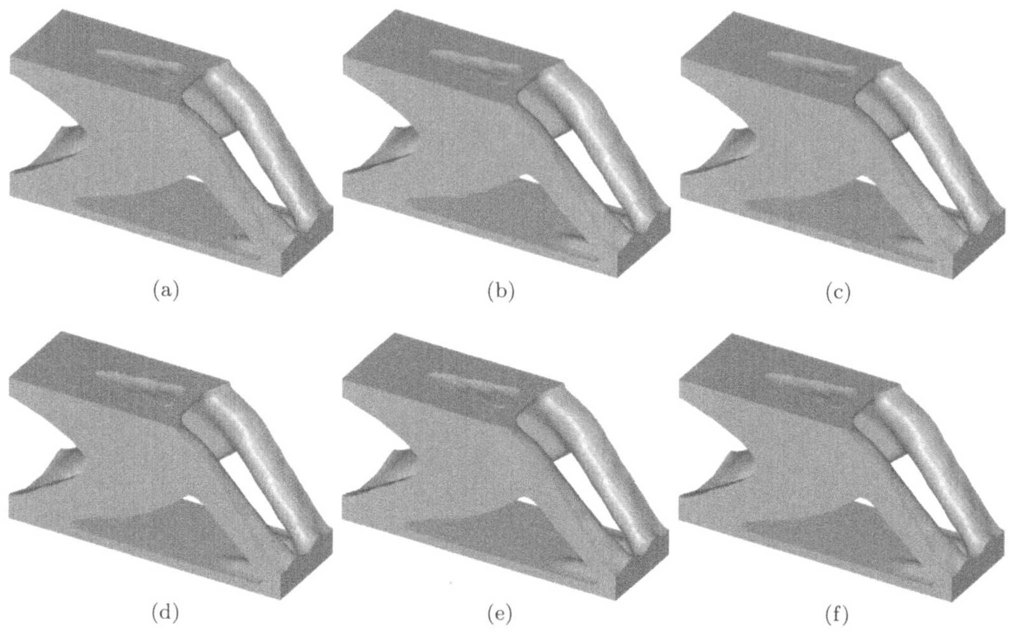

FIGURE 5.14
Resulting topologies of 3D short cantilever beam under different move limits in MMA: (a) $mov = 0.1$, (b) $mov = 0.35$, (c) $mov = 0.75$, (d) $mov = 1$, (e) $mov = 1.5$, and (f) $mov = 2$.

Figures 5.15 and 5.16, respectively. In Figure 5.15, the left side is fully fixed, and a unit vertical load ($F = -1$ N) is applied at the center point of the right side. In the 2D case, a mesh of 120×40 is utilized, V^* is set to 0.6, and $r_{\min}$ is set to 3. In Figure 5.16, the design domain is fully constrained at one end, and a unit vertical load ($F = -1$ N) is imposed in the middle of the free end. In the 3D case, a mesh of 48×16×16 is adopted, V^* is set to 0.3, and $r_{\min}$ is set to 1.5.

Six different numbers of grid points (5×5, 10×10, 15×15, 20×20, 25×25, and 30×30) in each element are explored for the 2D case, and the effects of the number of grid points on compliance, the number of iterations, and CPU time are shown in Figure 5.17. It can be seen that with the increase in the number of grid points, the CPU time increases dramatically. The influence of the number of grid points on structural compliance is not significant, and there is no clear relationship between the number of grid points and the number of iterations. When the number of grid points in each element is set to 25×25, the lowest compliance value

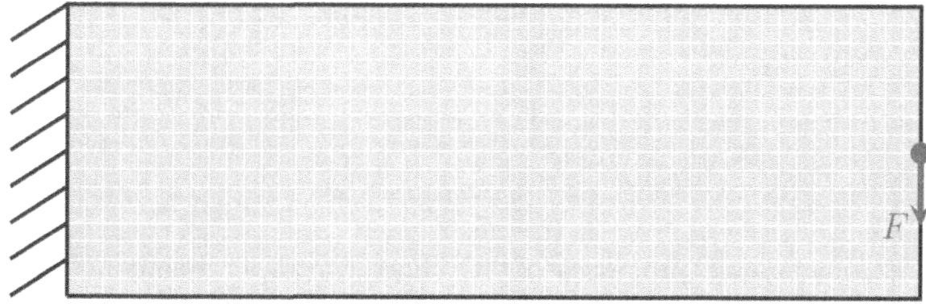

FIGURE 5.15
2D long cantilever beam.

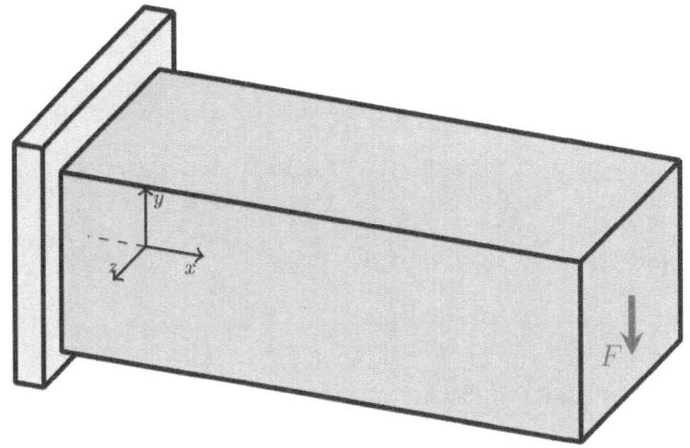

FIGURE 5.16
3D long cantilever beam.

is attained; however, the total number of iterations is the highest. The obtained topological designs under the six different numbers of grid points are shown in Figure 5.18. As shown in Figure 5.18, each set of grid points can lead to different topological configurations, and the topological design at 25×25 (Figure 5.18e) has more holes than other designs.

Four different numbers of grid points (2×2×2, 5×5×5, 8×8×8, and 10×10×10) in each element are considered for the 3D case, and the effects of the number of grid points on compliance, the number of iterations, and CPU time are shown in Figure 5.19. Similar with the 2D case, with the increase in the number of grid points, the CPU time increases significantly. The number of grid points has very slight effect on structural compliance, and there is no clear relationship between the number of grid points and the number of iterations. The obtained topological designs under the four different numbers of grid points

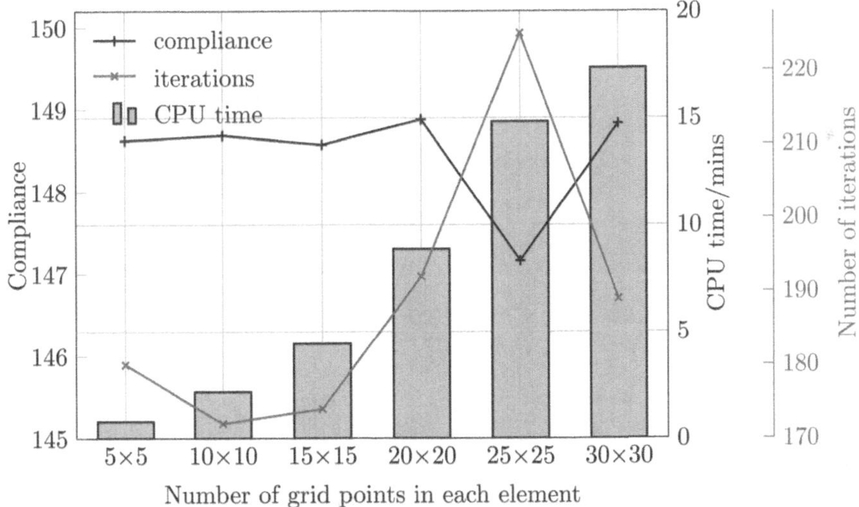

FIGURE 5.17
Effects of number of grid points on compliance, number of iterations, and CPU time for 2D case.

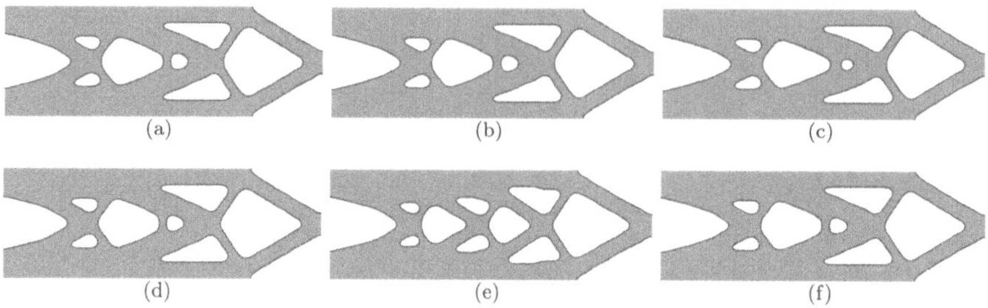

FIGURE 5.18
2D resulting topologies under different numbers of grid points: (a) 5×5, (b) 10×10, (c) 15×15, (d) 20×20, (e) 25×25, and (f) 30×30.

are shown in Figure 5.20. Although the four topological configurations are very similar, increasing the number of grid points in each element can improve the smoothness of the topological boundary.

It has been shown that the number of grid points in each element can affect the structural performance, the number of iterations, topological configuration, and boundary smoothness. The topological boundary is updated at each iteration of the optimization, and therefore it plays a role in the convergence of the optimization. Therefore, the smooth boundary strategy in SEMDOT cannot be regarded as a simple post-processing approach to topology optimization. In summary, the number of grid points in each element is set to 10×10 for the 2D case and 5×5×5 for the 3D case to maintain a proper balance between the computational efficiency and structural performance. Generally, improved solutions can be expected by conducting a parametric study on the number of grid points in each element.

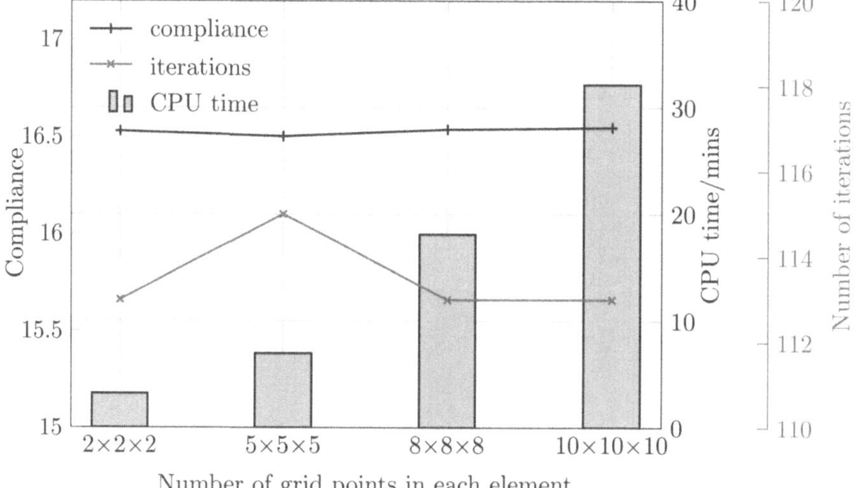

FIGURE 5.19
Effects of number of grid points on compliance, number of iterations, and CPU time for 3D case.

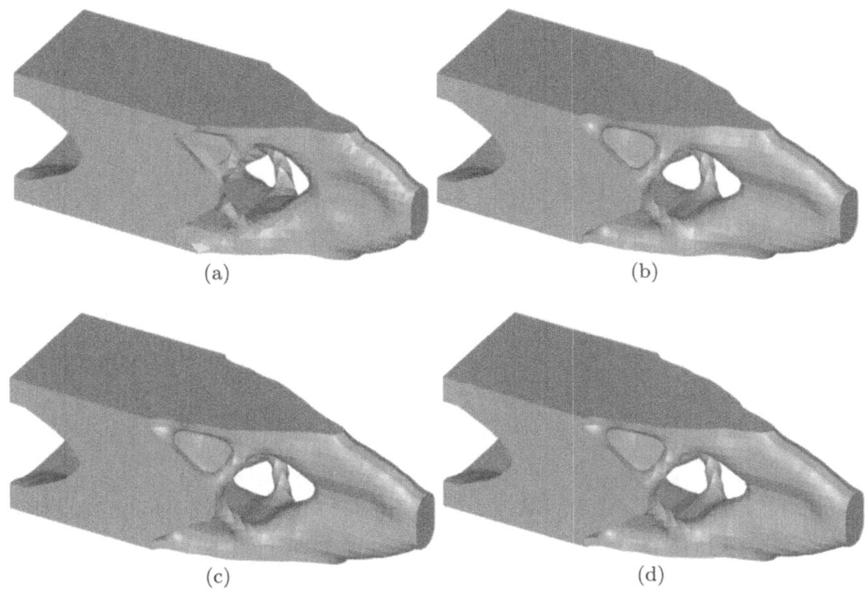

FIGURE 5.20
3D resulting topologies under different numbers of grid points: (a) 2×2×2, (b) 5×5×5, (c) 8×8×8, and (d) 10×10×10.

5.2.3 Effects of Filter Radius

The filter radius is an important parameter in element-based topology optimization algorithms. The 2D case in Figure 5.15 and 3D case in Figure 5.16 are employed to investigate the effects of the filter radius $r_{\min}$ on structural performance, convergence, and topological configuration. For the 2D case, a mesh of 360×120 is used, and V^* is set to 0.6. For the 3D case, a mesh of 90×30×20 is used, and V^* is set to 0.2.

The performance and convergence of the 2D optimization case under six different filter radii ($r_{\min}$ = 1.5, 2.5, 3.5, 4.5, 5.5, and 6.5) are summarized in Table 5.1. When $r_{\min} = 2.5$, the compliance is the lowest, which is 146.7152. After $r_{\min} = 2.5$, compliance goes up with the increase in $r_{\min}$. As can be seen in Table 5.1, the number of iterations rises with the increase in $r_{\min}$, meaning that a large $r_{\min}$ may prolong the convergence process. Resulting topologies under different filter radii are shown in Figure 5.21. It can be seen from Figure 5.21 that a small $r_{\min}$ can lead to a more complex geometry (with more holes).

Table 5.2 outlines the structural compliance values and numbers of iterations of the 3D case under six different filter radii, which are $r_{\min}$ = 1.5, 2.5, 3.5, 4.5, 5.5, and 6.5. With the increase in $r_{\min}$, the value of structural compliance rises. Although there is no clear relationship between the number of iterations and $r_{\min}$, a high value of $r_{\min}$ generally

TABLE 5.1
Compliance and convergence of 2D long cantilever beam under different filter radii

Filter radius $r_{\min}$	**1.5**	**2.5**	**3.5**	**4.5**	**5.5**	**6.5**
Compliance	146.7184	146.7152	146.7605	147.5714	147.5972	148.1384
Number of iterations	98	124	163	205	257	299

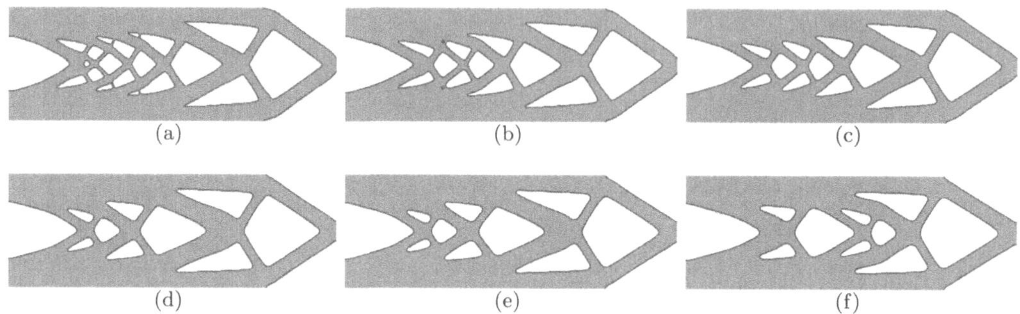

FIGURE 5.21
2D resulting topologies under different filter radii: (a) $r_{\min} = 1.5$, (b) $r_{\min} = 2.5$, (c) $r_{\min} = 3.5$, (d) $r_{\min} = 4.5$, (e) $r_{\min} = 5.5$, and (f) $r_{\min} = 6.5$.

results in a long convergence process. As can be seen in Figure 5.22, increasing the value of $r_{\min}$ is able to reduce the complexity of the topological design. A simple topological design may be preferred from the manufacturing point of view despite the fact that structural performance needs to be sacrificed.

As discussed in Chapter 2, if the filtering technique is not used in an element-based algorithm, such as SIMP, the checkerboard pattern may arise. However, in SEMDOT, the checkerboard pattern is less of an issue. When using $r_{\min} = 1$ in SEMDOT, the optimized 2D and 3D topological configurations are shown in Figure 5.23 where continuous structures can be observed. It is noted that the smoothness of topological designs shown in Figure 5.23 is not acceptable in comparison with the results obtained by $r_{\min} > 1$. Therefore, it is recommended to use $r_{\min} > 1$ in SEMDOT.

5.2.4 Multiple Load Cases

Multiple load cases are more common in structural optimization in comparison with single load cases. Both 2D and 3D cantilever beam cases are considered as shown in Figures 5.24 and 5.25, respectively. In Figure 5.24, the left side is fully fixed, and two unit vertical loads ($F_1 = -1$ N and $F_2 = 1$ N) are imposed at the bottom and top points of the right side, respectively. In the 2D case, a mesh of 200×100 is used, V^* is set to 0.5, and $r_{\min}$ is set to 3. As shown in Figure 5.25, the left end is fully fixed, and two vertical unit loads ($F_1 = -1$ N and $F_2 = 1$ N) are applied to the middle of the lower and upper edges of the free end, respectively. In the 3D case, a mesh of 80×40×26 is used, V^* is set to 0.3, and $r_{\min}$ is set to 2.5.

The objective function is the sum of multiple load cases, which is given by:

$$C(\tilde{\boldsymbol{X}}) = \sum_{i=1}^{N_{ml}} C_i(\tilde{\boldsymbol{X}}) = \sum_{i=1}^{N_{ml}} \mathbf{f}_i^{\mathrm{T}} \mathbf{u}_i(\tilde{\boldsymbol{X}}), \tag{5.5}$$

where $\tilde{\boldsymbol{X}}$ is the column vector form of filtered elemental volume fractions and N_{ml} is the number of load cases, in this case ($N_{ml} = 2$).

TABLE 5.2
Compliance and convergence of 3D long cantilever beam under different filter radii

Filter radius $r_{\min}$	**1.5**	**2.5**	**3.5**	**4.5**	**5.5**	**6.5**
Compliance	18.3870	19.2261	19.5496	20.6484	22.6010	24.2990
Number of iterations	129	192	742	568	417	379

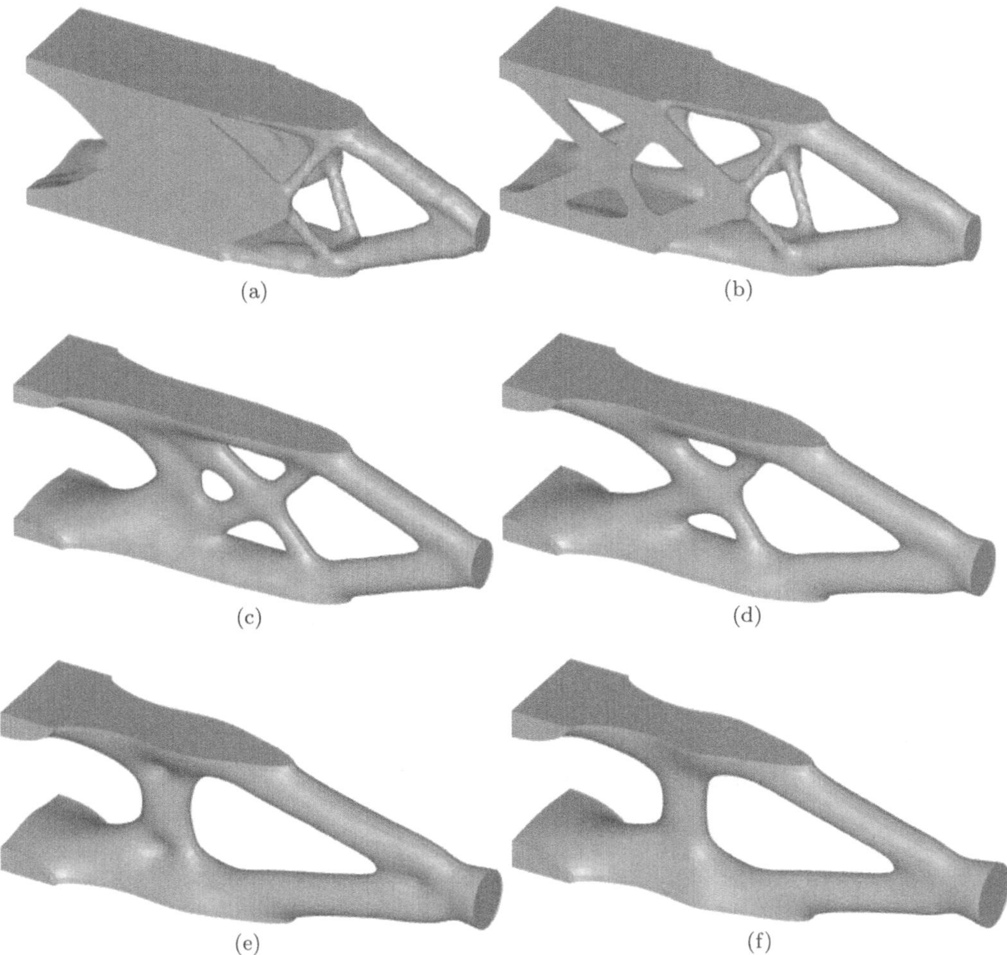

FIGURE 5.22
3D resulting topologies under different filter radii: (a) $r_{\text{min}} = 1.5$, (b) $r_{\text{min}} = 2.5$, (c) $r_{\text{min}} = 3.5$, (d) $r_{\text{min}} = 4.5$, (e) $r_{\text{min}} = 5.5$, and (f) $r_{\text{min}} = 6.5$.

The convergence process and optimized topology of the 2D cantilever beam with multiple loads are shown in Figure 5.26. The convergence process and optimized topology of the 3D cantilever beam with multiple loads are shown in Figure 5.27.

As can be seen in Figures 5.26 and 5.27, the optimization processes converge successfully, and 2D and 3D topological designs have some similarities.

5.2.5 Passive Elements

In some design problems, certain areas of the design domain need to remain void or solid. These are known as non-design regions or passive elements. Generally, void passive elements are given a value of 0 to define void areas, and solid passive elements are given a value of 1 to define solid or non-design regions. However, because of using the filtering technique that calculates the average value within a circular neighborhood with a radius of r_{min}, the elemental values in the design domain can be disturbed, which may adversely affect the beginning of optimization.

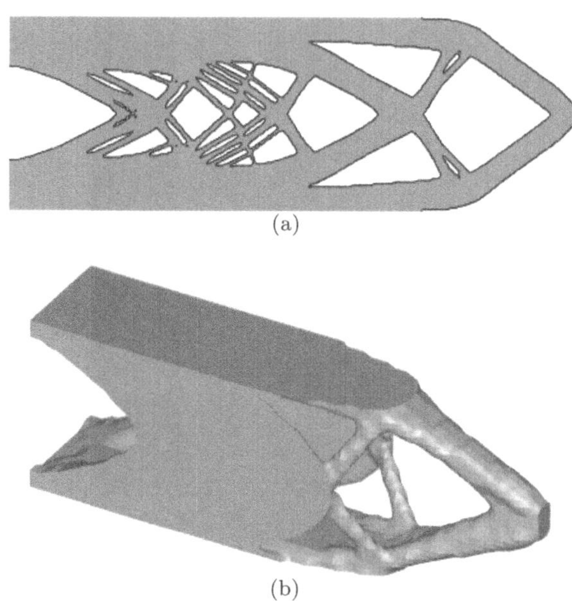

FIGURE 5.23
Resulting topologies with $r_{\min} = 1$: (a) 2D and (b) 3D cases.

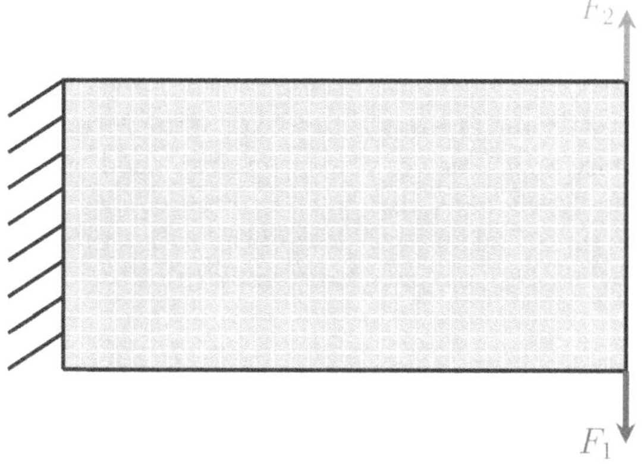

FIGURE 5.24
2D cantilever beam with multiple loads.

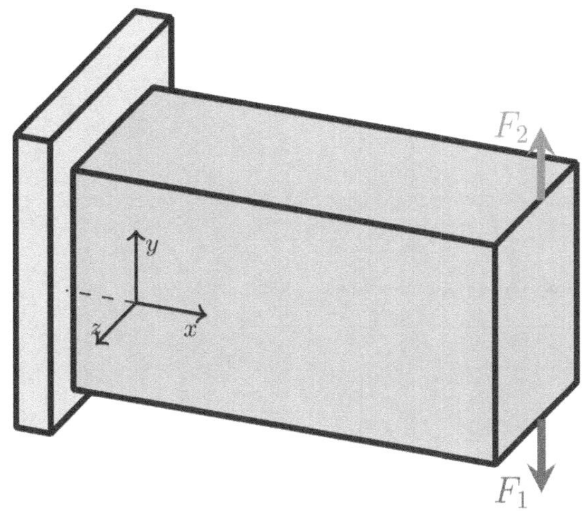

FIGURE 5.25
3D cantilever beam with multiple loads.

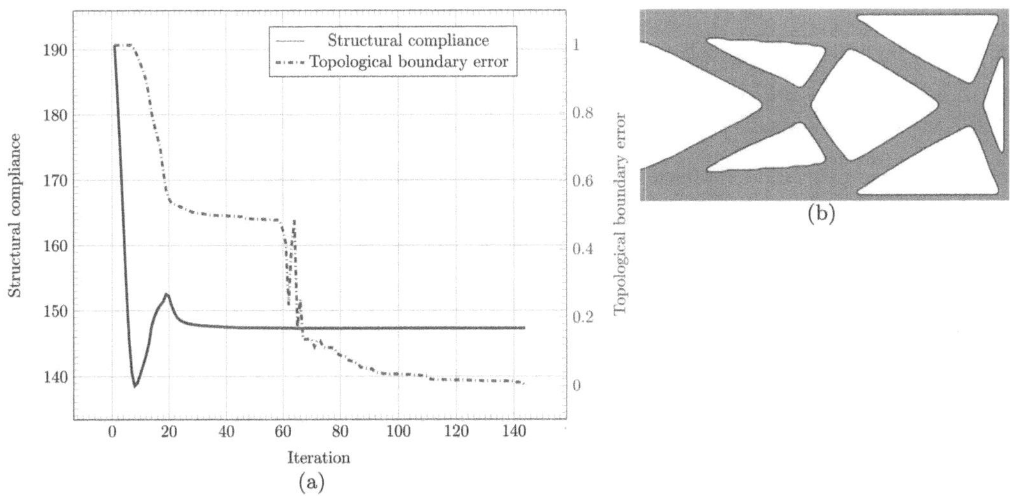

FIGURE 5.26
2D cantilever beam with multiple loads: (a) convergence process and (b) optimized topology.

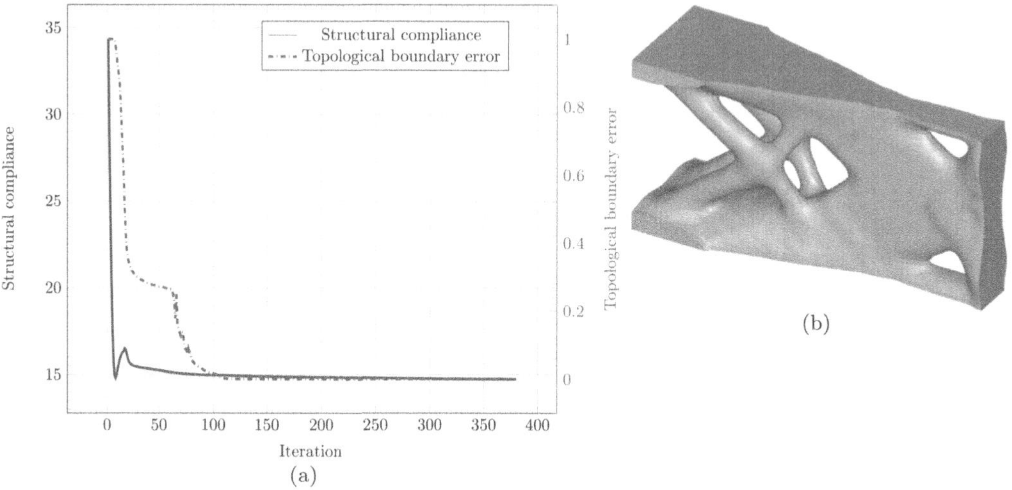

FIGURE 5.27
3D cantilever beam with multiple loads: (a) convergence process and (b) optimized topology.

The 2D cantilever beam case with a void region is shown in Figure 5.28. In this case, a mesh of 150×100 is adopted, V^* is set to 0.5, and r_{min} is set to 5. A circular region of the design domain with a radius of one third of the height of the design domain is fixed to be void. The center of the circular void region is at the half of the height and one third of the length of the design domain. The convergence process and optimized topology are shown in Figure 5.29.

The 3D cantilever beam case with a void region is shown in Figure 5.30. In this case, a mesh of 60×20×10 is adopted, V^* is set to 0.5, and r_{min} is set to 1.5. A cylindrical region of the design domain with a radius of one third of the height of the design domain is fixed to be a void. The center of the cylindrical void region is at the half of the height, the half of the width, and one third of the length of the design domain. The convergence process and optimized topology are shown in Figure 5.31.

The obvious oscillations can be observed at the beginning of the optimization processes in Figures 5.29(a) and 5.31(a). This is because the passive void elements interfere with the

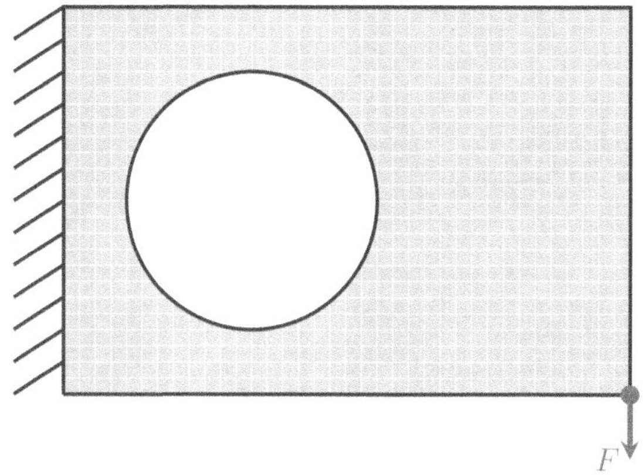

FIGURE 5.28
2D cantilever beam with void region.

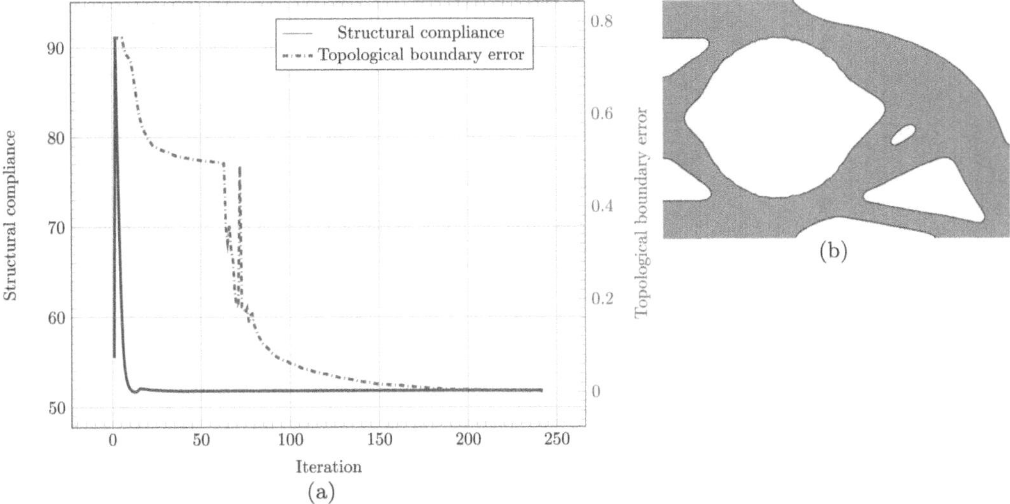

FIGURE 5.29
2D cantilever beam with void region: (a) convergence process and (b) optimized topology.

initial optimization process. The prescribed passive void areas can be clearly observed in the optimized topologies in Figures 5.29(b) and 5.31(b).

The 2D cantilever beam case with a solid non-design region is shown in Figure 5.32, where the circular area is the non-design region. It is noted that the parameter settings remains unchanged from the previous respective examples. The convergence process and optimized topology are shown in Figure 5.33.

The 3D cantilever beam case with a solid region is shown in Figure 5.34, where the cylindrical region is the non-design area. The parameter settings still remain the same as the previous respective examples. The convergence process and optimized topology are shown in Figure 5.35.

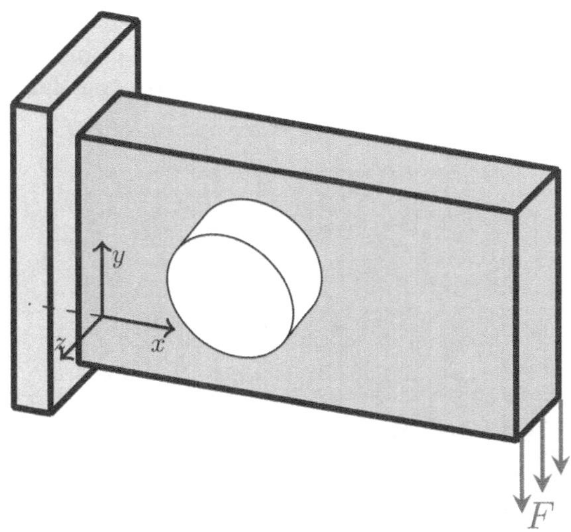

FIGURE 5.30
3D cantilever beam with void region.

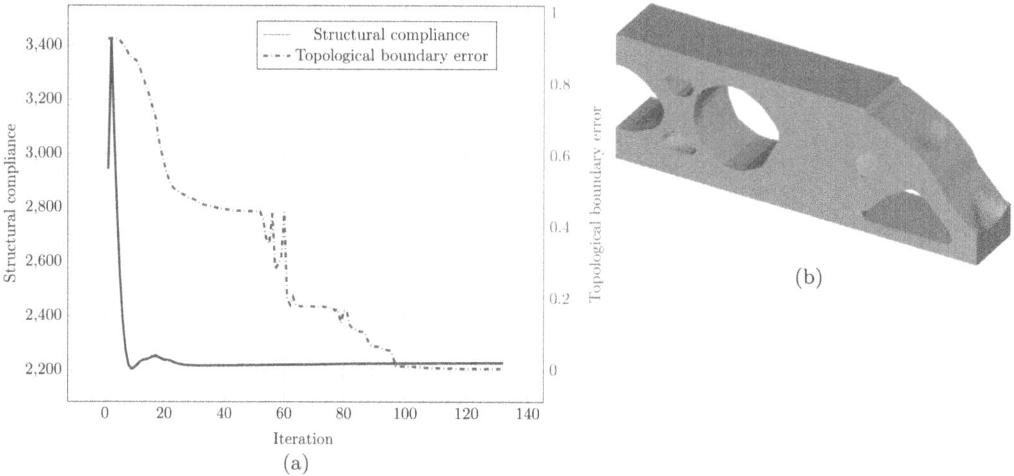

FIGURE 5.31
3D cantilever beam with void region: (a) convergence process and (b) optimized topology.

It has been shown, similar to the cases with passive element void areas, that obvious oscillations are observed at the beginning of the optimization processes in Figures 5.33(a) and 5.35(a) since the passive solid elements interfere with the initial optimization process. The predefined solid or non-design areas can be clearly observed in the topological designs in Figures 5.33(b) and 5.35(b).

5.2.6 Real-World Optimization Cases

Although the effectiveness of SEMDOT has been validated by standard 2D and 3D benchmark optimization problems in previous sections, its capability in solving real-world cases must also be proven.

5.2.6.1 Sheikh-Ibrahim's Steel Beam Splice Connection

The first real-world engineering case is the topological design of the Sheikh-Ibrahim's steel beam splice connection. The splice connection layout is shown in Figure 5.36, where the dimensions are given. The design domain and loads of the splice connection are shown in Figure 5.37. In Figure 5.37, passive void elements are used to define holes, and passive solid

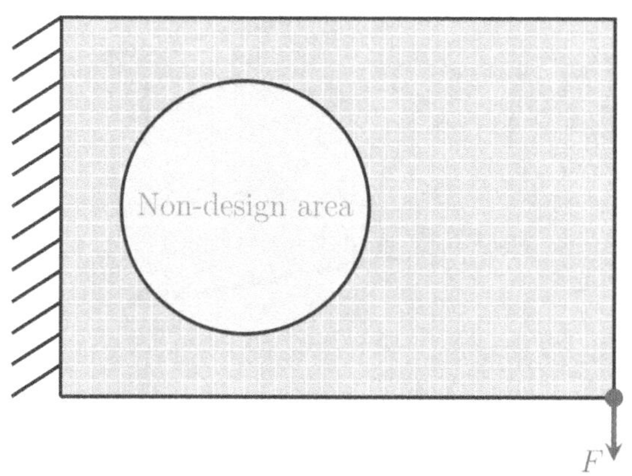

FIGURE 5.32
2D cantilever beam with solid region.

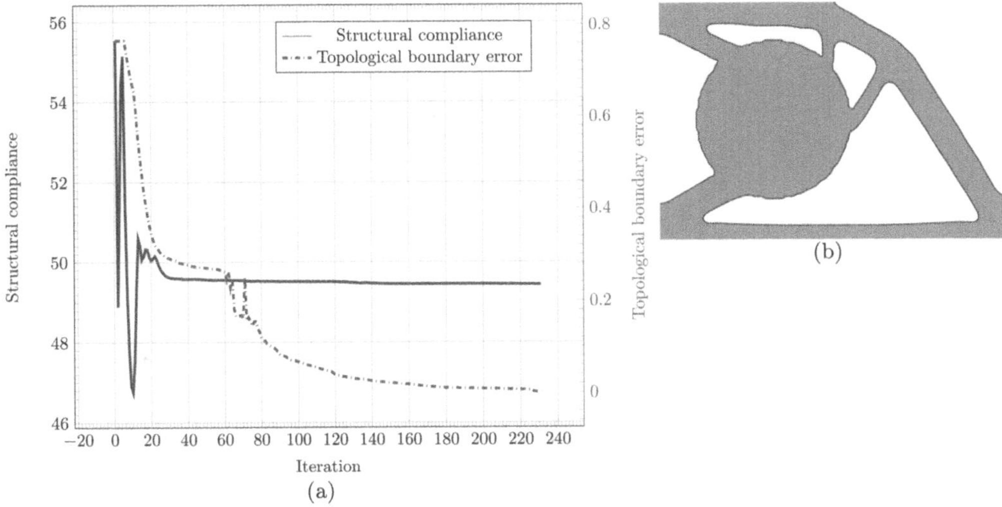

FIGURE 5.33
2D cantilever beam with solid region: (a) convergence process and (b) optimized topology.

elements are used to define non-design areas. The diameter of the non-design area is 49.7. In SEMDOT, a mesh of 814×206 is employed, $r_{\min}$ is set to 2 and V^* is set to 0.55. Here, the evolution rate Λ is set to 0.25.

The convergence process and optimized topology of the splice connection are shown in Figure 5.38. Due to the use of both passive solid and void elements, an oscillation is found at the beginning of the optimization process (see Figure 5.38(a)). The optimized topology presented in Figure 5.38(a) can be easily fabricated by additive manufacturing because of the smooth boundary generated by SEMDOT. The printed structure is shown in Figure 5.39. More details on this engineering case can be found in [149–151].

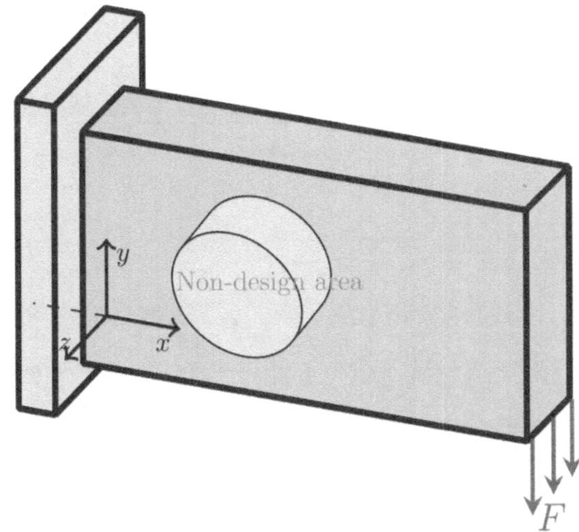

FIGURE 5.34
3D cantilever beam with solid region.

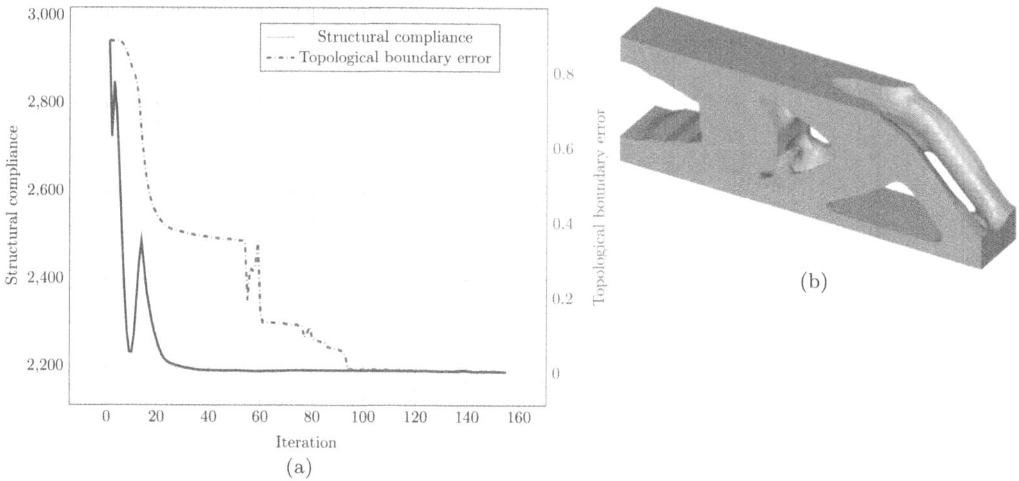

FIGURE 5.35
3D cantilever beam with solid region: (a) convergence process and (b) optimized topology.

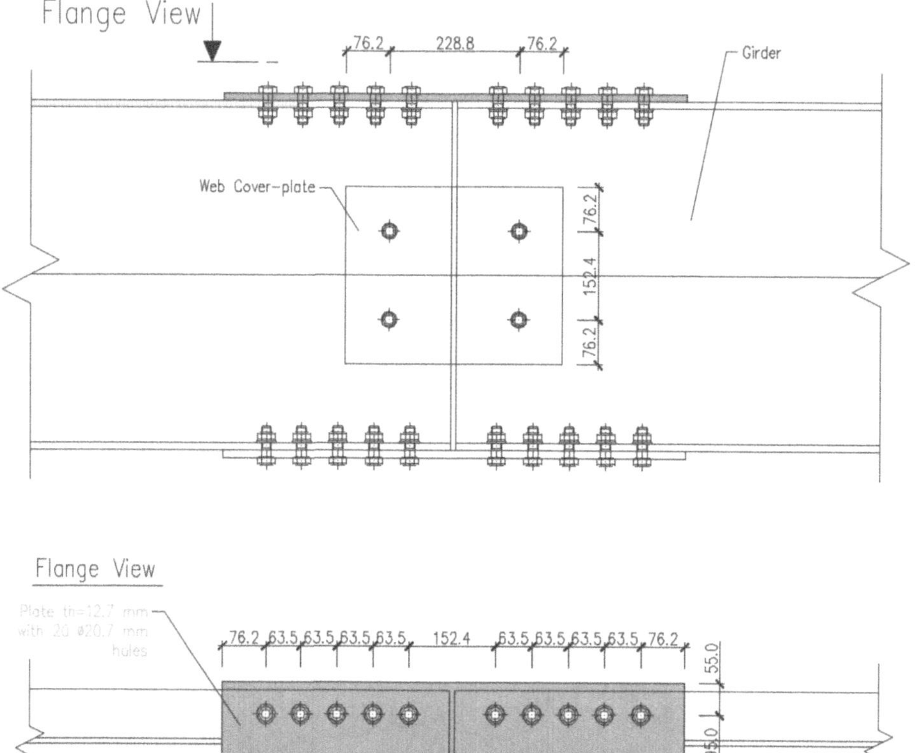

FIGURE 5.36
Splice connection layout [151].

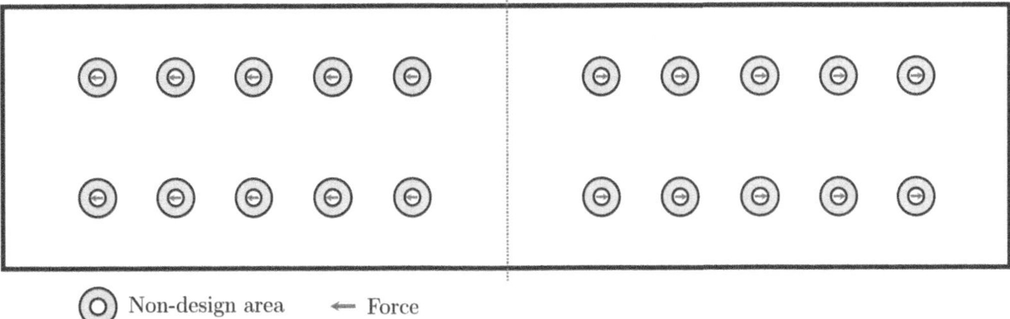

FIGURE 5.37
Design domain of splice connection.

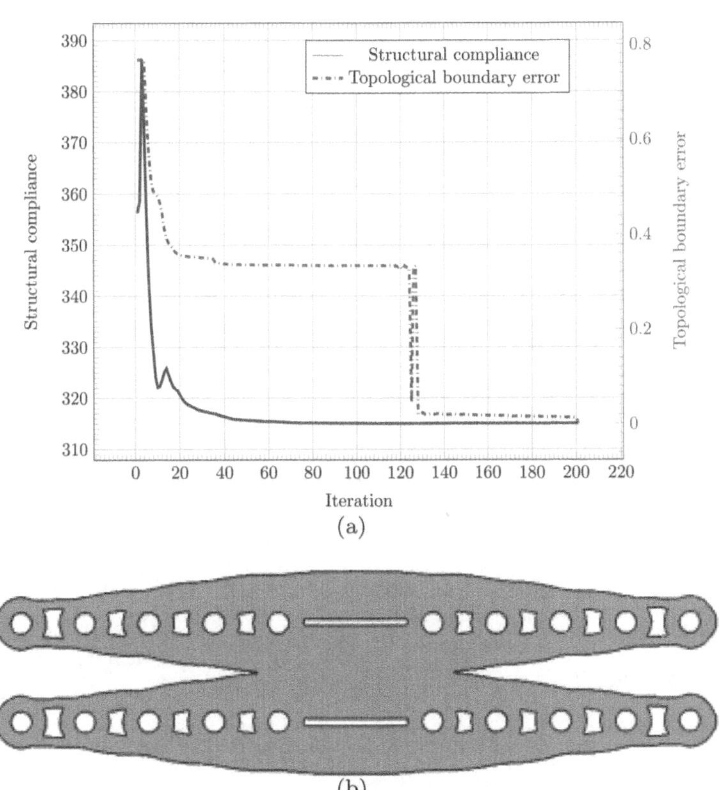

FIGURE 5.38
Splice connection case: (a) convergence process and (b) optimized topology.

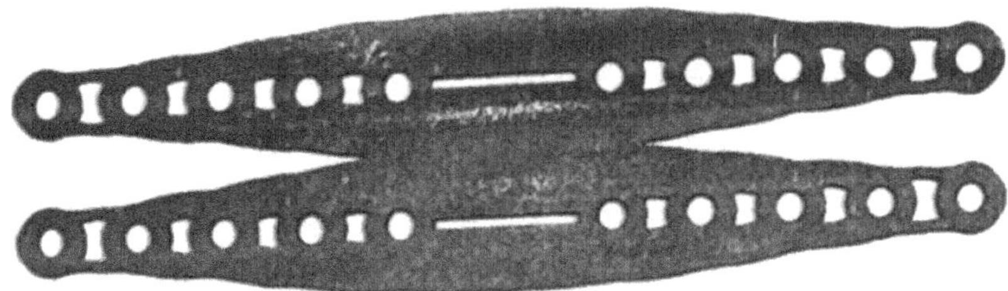

FIGURE 5.39
Additively manufactured splice connection.

5.2.6.2 Suspension Support Rod of Vehicle Rear Axle

The second real-world engineering case is the topological design of the suspension support rod of the vehicle rear axle. The design domain, loading, and boundary conditions are shown in Figure 5.40, where the torque is imposed on the two loading areas, and the fully fixed constraint is applied to the boundary area. Here, a mesh of 40×20×60 is used, $r_{\min}$ is set to 3, and V^* is set to 0.2.

The convergence process, initial design, and optimized topology of the suspension support rod are shown in Figure 5.41. Because there are non-design regions in the optimization problem, obvious oscillations can be found at the beginning of optimization (see Figure 5.41(a)). After the initial oscillation, the optimization process gradually converges toward a constant value, meaning that optimization converges successfully. The initial design and optimized topology are shown in Figure 5.41(b and c), respectively. To clearly show the geometry, the topologically optimized suspension support rod seen from various orientations is shown in Figure 5.42. The additively manufactured part is shown with support structures and then after the removal of the support structures in Figure 5.43.

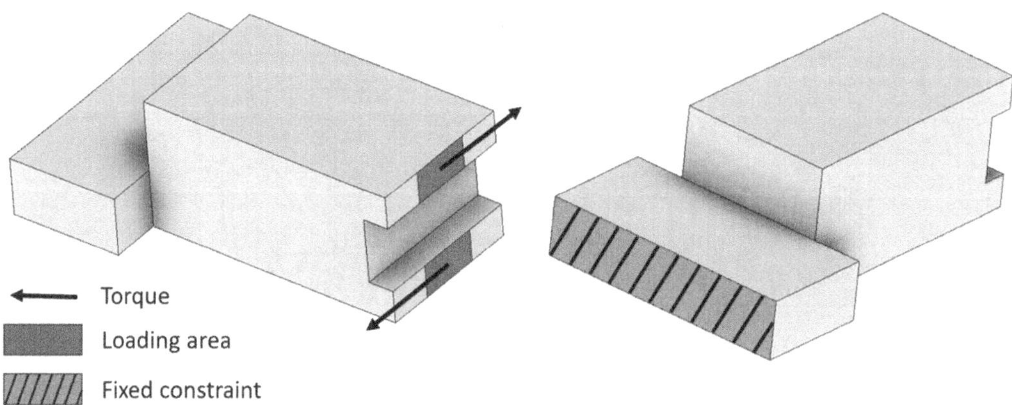

FIGURE 5.40
Loading and boundary conditions of suspension support rod.

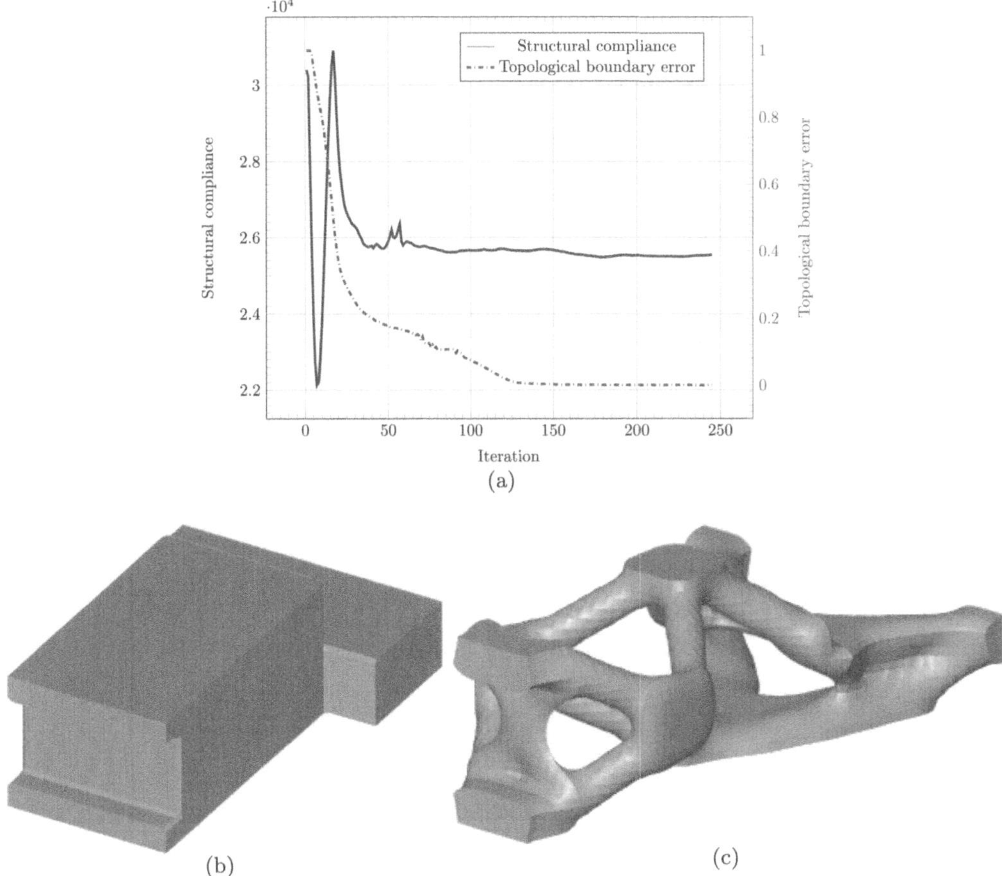

FIGURE 5.41
Suspension support rod: (a) convergence process, (b) initial design, and (c) optimized topology.

5.3 Summary

This chapter investigated the effect of a number of variables on the optimization of compliance minimization problems. The move limit in OC and MMA optimizers has very little effect on the final results when solving the classic stiffness maximization problems. In general, MMA can obtain better solutions compared to OC. Therefore, the MMA optimizer is used as the default optimization solver for SEMDOT, and will be used in the subsequent chapters unless otherwise instructed. Increasing the number of grid points can improve the smoothness of topological boundaries; however, this significantly increases the computational cost. Changing the number of grid points can yield different topological configurations, and hence, the smooth boundary algorithm strategy in SEMDOT cannot be simply regarded as a post-processing method. Increasing the value of the filter radius can reduce the complexity of the optimized topology; however, the structural performance will be sacrificed. No checkerboard pattern is observed when the filter radius is set to 1 in SEMDOT; however, other complications may occur when the filter radius is equal to 1. Multiple loads and non-design regions can be easily defined in SEMDOT. Most importantly, SEMDOT can

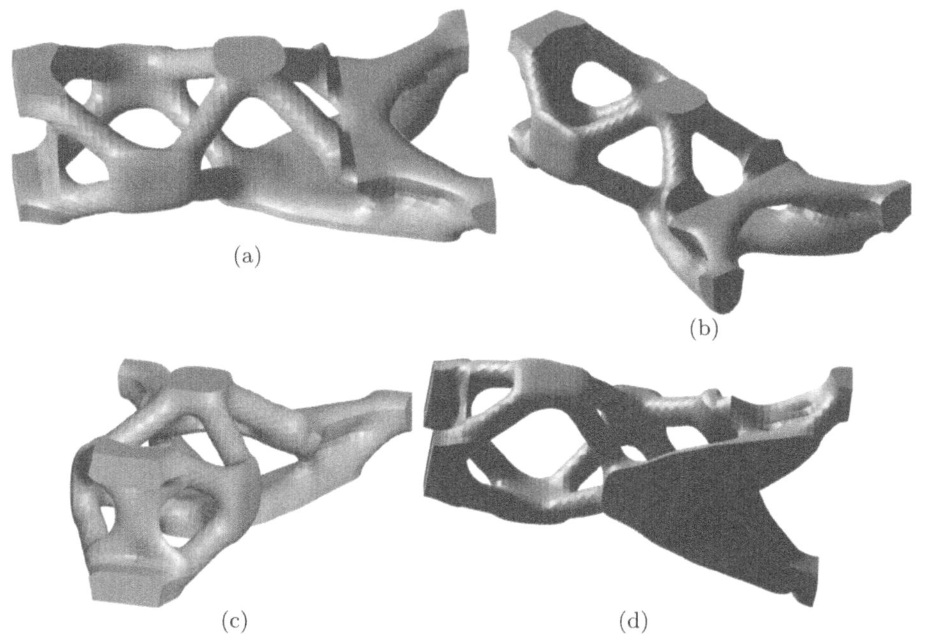

(a) (b) (c) (d)

FIGURE 5.42
Topologically optimized suspension support rod under different orientations.

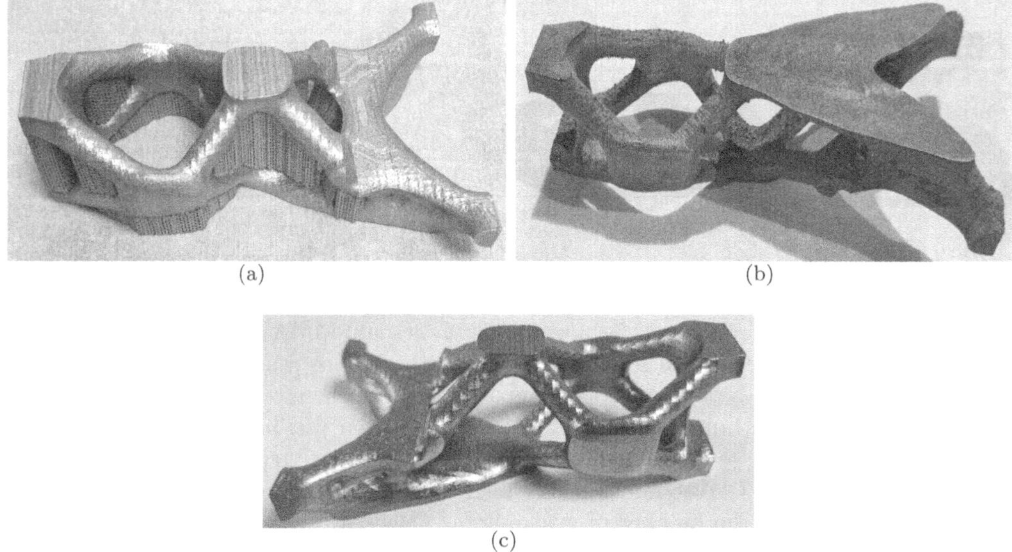

(a) (b) (c)

FIGURE 5.43
Additively manufactured suspension support rod (a) with support structures, (b) after removing support structures, and (c) after polishing.

be used to deal with real-world optimization problems, for example, Sheikh-Ibrahim's steel beam splice connection and suspension support rod of vehicle rear axle. It can be concluded that SEMDOT is an effective algorithm for solving stiffness maximization/compliance minimization problems.

6

Compliant Mechanism Design Problems

This chapter uses the Smooth-Edged Material Distribution for Optimizing Topology (SEMDOT) algorithm to solve compliant mechanism design problems. In this chapter, the effects of varying the move limit in the Method of Moving Asymptote (MMA) optimizer, and the input and output spring stiffnesses on the final results (performance, convergence, and configurations) are systematically discussed through both 2D and 3D cases.

6.1 Optimization Problem

According to Zhu et al. [220], compliant mechanisms are an important branch of modern mechanisms. In contrast to traditional rigid body mechanisms, compliant mechanisms are able to convert displacement and loads into deformation of the mechanisms structure. This deformation of structure can lead to reductions in the need for lubrication or assembly [220]. A generic design domain for compliant mechanisms with a single input-output scenario is illustrated in Figure 6.1. An important application of compliant mechanisms is that of MicroElectroMechanical Systems (MEMS). MEMS cannot be easily fabricated using typical assembly processes and they struggle to make use of hinges and bearings since friction dominates the systems at the small scale [20].

Topology optimization has become an important tool to design compliant mechanisms [77, 84, 157]. Topology optimization of compliant mechanisms can be conducted on the basis of either continuum or truss and frame discretizations. This chapter concentrates on the continuum discretization option. Generally, the design of a compliant mechanism is regarded as a more challenging optimization problem compared to the stiffness maximization case. The increased difficulty of design of a compliant mechanism is the dynamic nature of the problem, where the mechanism's motion is given by the deformation of the structure. Though the stiffness maximization is a static load situation, the objective of the compliant mechanism optimization is to achieve the maximum output displacement u_{out}. The corresponding optimization problem is stated as:

$$\begin{aligned} \min : C(X_e) &= -u_{out} = -\mathbf{L}^{\mathbf{T}}\mathbf{u} = \tilde{\mathbf{u}}^{\mathbf{T}}\mathbf{K}\mathbf{u} \\ \text{subject to} : \mathbf{K}(X_e)\mathbf{u} &= \mathbf{f}_{in}, \\ &\frac{\sum\limits_{e=1}^{M} X_e V_e}{\sum\limits_{e=1}^{M} V_e} - V^* \leq 0, \\ &0 < \rho_{\min} \leq X_e \leq 1;\ e = 1, 2, \cdots, M, \end{aligned} \tag{6.1}$$

DOI: 10.1201/9781032634449-6

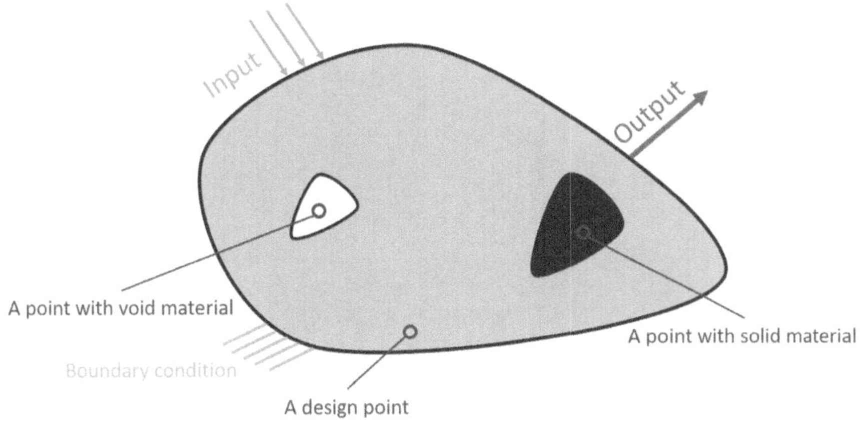

FIGURE 6.1
Compliant mechanisms with a single input-output.

where $C(X_e)$ is the objective function, X_e is the elemental volume fraction, $\mathbf{L}$ is a unit length vector with zeros at all degrees of freedom except at the output point where it is one, $\mathbf{u}$ is the displacement vector, $\tilde{\mathbf{u}}$ is the dummy load displacement vector calculated by solving the adjoint problem $\mathbf{K}\tilde{\mathbf{u}} = -\mathbf{L}$, $\mathbf{f}_{in}$ is the input force vector, M is the total number of elements, V_e is the volume of the eth element, V^* is the target volume, and $\rho_{\min}$ is a small value, which the authors suggest should be 0.001 for this example.

Sensitivities of solid and void elements are [112] :

$$\frac{\partial C(X_e)}{\partial X_e} = \begin{cases} \dfrac{C(X_e=1) - C(X_e=0)}{1} \approx \tilde{\mathbf{u}}_{\mathbf{e}}^{\mathbf{T}} \mathbf{K}_e^1 \mathbf{u}_e & \text{if } X_e = 1, \\ \dfrac{C(X_e=0) - C(X_e=1)}{-1} \approx \rho_{\min} \tilde{\mathbf{u}}_{\mathbf{e}}^{\mathbf{T}} \mathbf{K}_e^1 \mathbf{u}_e & \text{if } X_e = \rho_{\min}, \end{cases} \tag{6.2}$$

where $\mathbf{K}_e^1$ is the stiffness matrix of the solid element.

Sensitivities of boundary elements are approximated as:

$$\begin{aligned} \frac{\partial C(X_e)}{\partial X_e} \approx &(1 - X_e) \left.\frac{\partial C(X_e)}{\partial X_e}\right|_{X_e=\rho_{\min}} + X_e \left.\frac{\partial C(X_e)}{\partial X_e}\right|_{X_e=1} \\ = &\, [(1 - X_e)\rho_{\min} + X_e] \tilde{\mathbf{u}}_e^{\mathbf{T}} \mathbf{K}_e^1 \mathbf{u}_e. \end{aligned} \tag{6.3}$$

6.2 Numerical Examples

6.2.1 Effects of Varying the Move Limit in MMA

For the 2D case, the compliant mechanism design is shown in Figure 6.2, where a single-piece flexible structure transfers an input force to another point through elastic deformation [77,157]. In Figure 6.2, f_{in} represents the input force; k_{in} and k_{out} represent input and output spring stiffnesses, respectively; and u_{out} represents the output displacement. For the 3D case, the compliant mechanism design is shown in Figure 6.3, where an input load is defined in the positive direction, and both the top and side faces have symmetry constraints, meaning that the constrained nodes can only move within the plane. The goal is to maximize the horizontal output displacement in the negative direction.

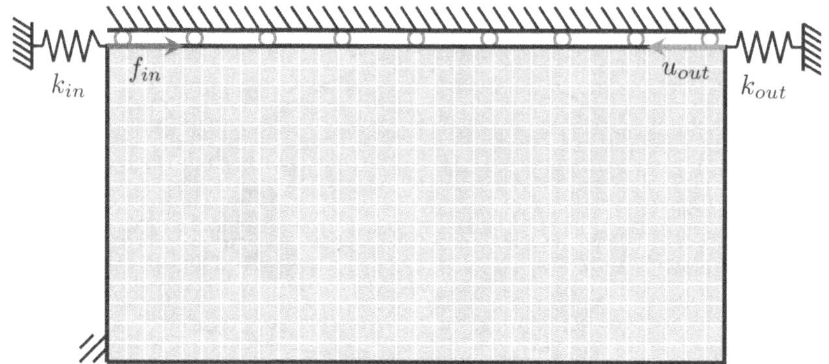

FIGURE 6.2
2D force inverter design problem.

Here, the effects of the move limit in Method of Moving Asymptote (MMA) on the performance and convergence are studied using both the 2D and 3D cases. In the 2D case, a mesh of 60×30, $f_{\text{in}} = 1$, $k_{\text{in}} = 0.1$, $k_{\text{out}} = 0.1$, $r_{\text{min}} = 1.5$, and $V^* = 0.3$ are applied. In the 3D case, a mesh of $70 \times 35 \times 6$, $k_{\text{in}} = 0.1$, $k_{\text{out}} = 0.1$, $r_{\text{min}} = 2$, and $V^* = 0.3$ are adopted.

When using the move limit of 1 ($mov = 1$), the convergence process and optimized topology for the 2D case are shown in Figure 6.4. The optimization procedure converges after 99 iterations at the output displacement of 0.9986. The output displacement and convergence of the 2D force inverter design case under different move limits (0.1, 0.5, 0.75, 1.5, 2, and 2.5) are summarized in Table 6.1. As outlined in Table 6.1, the lowest move limit ($mov = 0.1$) leads to the highest value of the output displacement (0.9915); however, the number of iterations is also the highest, which is 135. The optimized topologies under different move limits are shown in Figure 6.4, where similar topological configurations can be observed. Overall, the effects of the move limit on the output displacement and number of iterations for the 2D case are not significant.

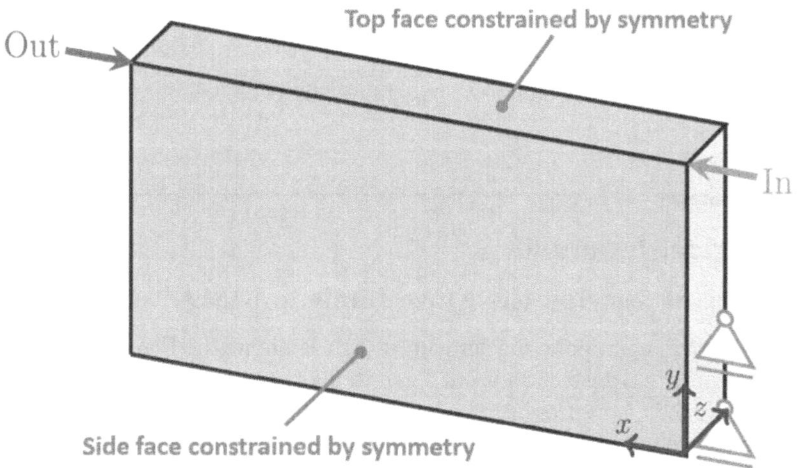

FIGURE 6.3
3D force inverter design problem.

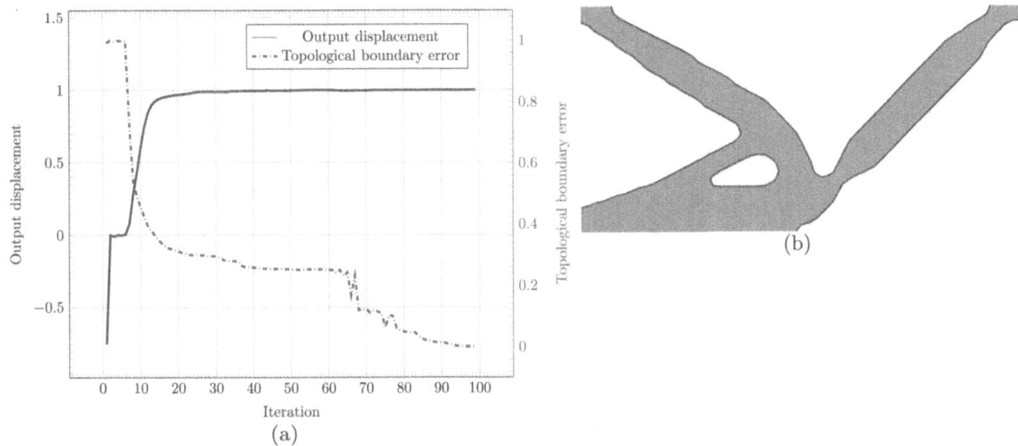

FIGURE 6.4
2D force inverter design case: (a) convergence process and (b) optimized topology.

TABLE 6.1
Output displacement and convergence of 2D force inverter design case under different move limits.

Move limit *mov*	**0.1**	**0.5**	**0.75**	**1.5**	**2**	**2.5**
Output displacement	0.9915	0.9988	0.9987	0.9986	0.9986	0.9986
Number of iterations	135	107	101	99	99	99

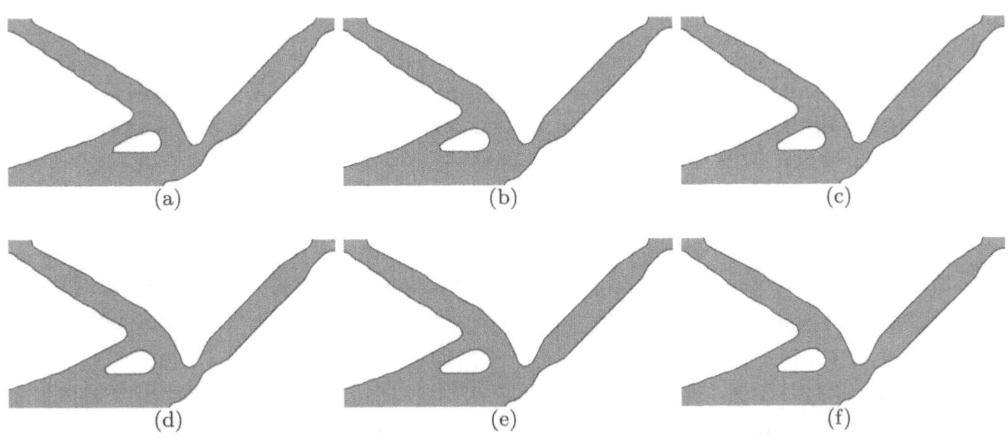

FIGURE 6.5
Resulting topologies of 2D force inverter design under different move limits: (a) $mov = 0.1$, (b) $mov = 0.5$, (c) $mov = 0.75$, (d) $mov = 1.5$, (e) $mov = 2$, and (f) $mov = 2.5$.

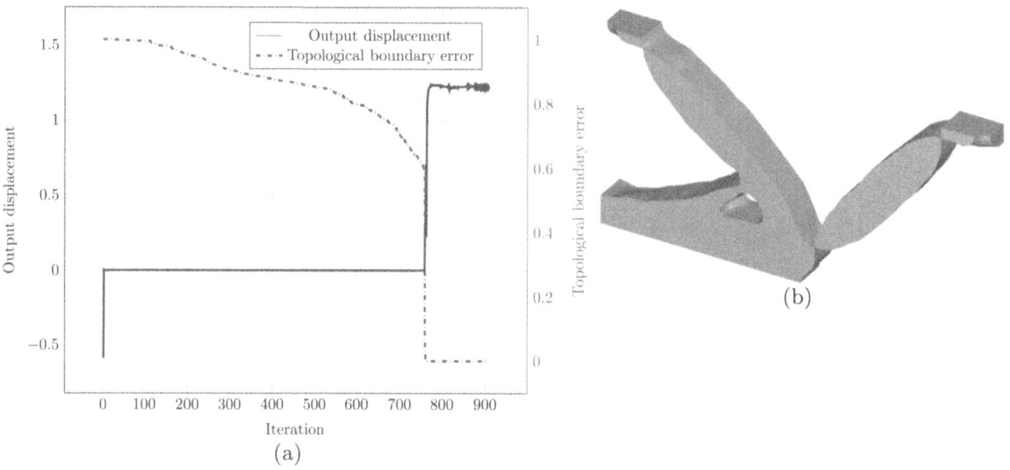

FIGURE 6.6
3D force inverter design case: (a) convergence process and (b) optimized topology.

When using $mov = 1$, the convergence process and optimized topology for the 3D case are shown in Figure 6.6. The optimization procedure converges after 908 iterations at the output displacement of 1.2238. The output displacement and convergence of the 3D force inverter design case under different move limits (0.1, 0.25, 0.5, 0.75, 1.5, and 2) are summarized in Table 6.2. In contrast to the 2D case, the high move limit (for example, $mov = 1.5$ and 2) can contribute to the highest output displacement with a value of 1.2238; however, a high number of iterations (908) is required. The move limit of 0.25 ($mov = 0.25$) leads to the lowest number of iterations, which is 248. When using the move limits of 0.5 and 0.75, the output displacement is determined to be zero, and hence these two optimization processes are not successful. The optimized topologies under different move limits are shown in Figure 6.7. Obviously, different move limits lead to distinct topological configurations, and there are two failed structures when using the move limits of 0.5 and 0.75 (see Figure 6.7c,d). Compared to the 2D case, the 3D case is much more sensitive to the move limit.

To maintain a proper balance between performance and convergence for both 2D and 3D cases, the move limit of 0.25 ($mov = 0.25$) is recommended by the authors, and this is used for the rest of the chapter. Generally, a small move limit in MMA can stabilize the convergence of optimization. The selection of the move limit should be based on each specific case. In some cases, the convergence time has to be sacrificed to obtain better performance.

6.2.2 Effects of Varying the Input Spring Stiffness

The influences of varying the input spring stiffnesses (k_{in}=0.1, 0.01, 0.001, and 0.0001) on the output displacement and convergence period are investigated using both the 2D

TABLE 6.2
Output displacement and convergence of 3D force inverter design case under different move limits.

Move limit mov	**0.1**	**0.25**	**0.5**	**0.75**	**1.5**	**2**
Output displacement	1.1766	1.2025	0	0	1.2238	1.2238
Number of iterations	308	248	883	1998	908	908

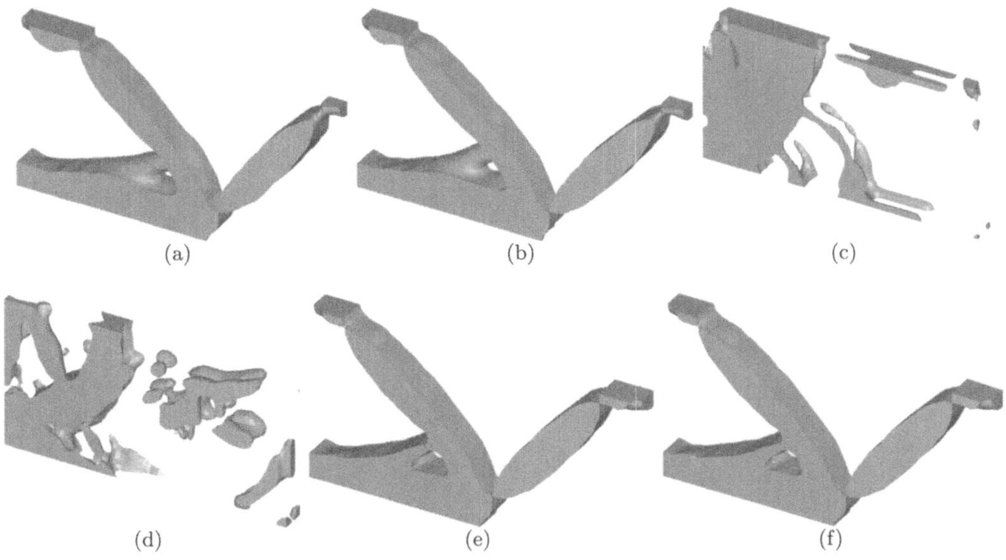

FIGURE 6.7
Resulting topologies of 3D force inverter design under different move limits: (a) $mov = 0.1$, (b) $mov = 0.25$, (c) $mov = 0.5$, (d) $mov = 0.75$, (e) $mov = 1.5$, and (f) $mov = 2$.

and 3D cases presented in Figures 6.2 and 6.3. For the 2D case, a mesh of 240×120, $k_{out} = 1$, $r_{min} = 4$, and $V^* = 0.3$ are adopted. The output displacement and number of iterations of the 2D force inverter design case under different input spring stiffnesses are summarized in Table 6.3. As outlined in Table 6.3, when $k_{in} = 0.1$, the convergence is slowest and the number of iterations is the largest (319). With the decrease in k_{in}, the output displacement increases. The resulting final configurations of the 2D force inverter design case under the various input spring stiffnesses are shown in Figure 6.8. It can be seen from Figure 6.8 that the higher values of k_{in} result in thinner features connected to the input actuator. The lower values of k_{in} lead to thinner features connected to the output actuator. This diversity in configurations does lead to an increase in complexity when fabricating the structure. It should also be noted that the structures do not break or fail during the analysis.

For the 3D case, a mesh of $120 \times 60 \times 8$, $k_{out} = 1$, $r_{min} = 2.5$, and $V^* = 0.3$ are used. The output displacement and convergence of the 3D force inverter design case under different input spring stiffnesses are outlined in Table 6.4. In Table 6.4, when $k_{in} = 0.1$, the number of iterations is the largest (1,190). In addition, when the k_{in} decreases, then the output displacement increases. The resulting final configurations of the 3D force inverter design case under the different input spring stiffnesses are shown in Figure 6.9. The changes in

TABLE 6.3
Output displacement and convergence of 2D force inverter design case under different input spring stiffnesses.

Input spring stiffness k_{in}	**0.1**	**0.01**	**0.001**	**0.0001**
Output displacement	0.1017	0.5529	1.5097	2.1033
Number of iterations	319	159	213	245

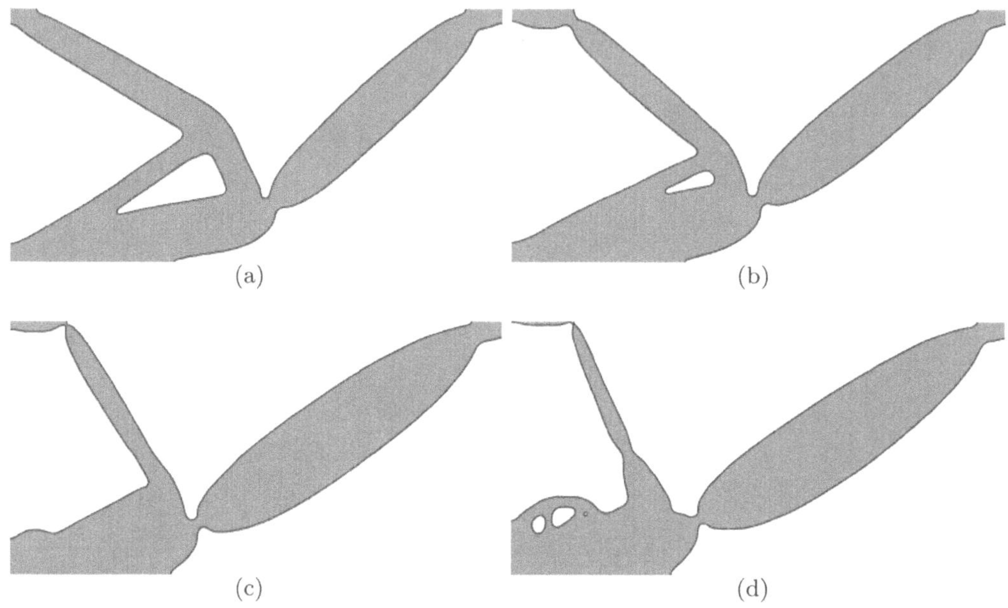

FIGURE 6.8
Resulting topologies of 2D force inverter design case under different input spring stiffnesses: (a) k_{in}=0.1, (b) k_{in}=0.01, (c) k_{in}=0.001, and (d) k_{in}=0.0001.

the input spring stiffness, k_{in}, lead to very different 3D final topological configurations. It is not clear if there is any trend on the mechanism features when the input spring stiffness is varied in the 3D case.

6.2.3 Effects of Varying the Output Spring Stiffness

The influences of varying the output spring stiffnesses (k_{out}=0.1, 0.01, 0.001, and 0.0001) on the output displacement and convergence period are also studied using both the 2D and 3D cases presented in Figures 6.2 and 6.3. The parameter settings are the same as with Section 6.2.2. The output displacement and number of iterations of the 2D force inverter design case under the various output spring stiffnesses are summarized in Table 6.5, where when $k_{\text{out}} = 0.1$, the convergence period is longest and the number of iterations is the largest (480). It is also seen that when output spring stiffness, k_{out}, decreases, then the output displacement increases. The resulting topologies configurations of the 2D force inverter design case under the various output spring stiffnesses are shown in Figure 6.10. It can be seen from Figure 6.10, that varying the output spring stiffness, k_{out}, leads to

TABLE 6.4
Output displacement and convergence of 3D force inverter design case under different input spring stiffnesses

Input spring stiffness k_{in}	**0.1**	**0.01**	**0.001**	**0.0001**
Output displacement	0.1604	0.5446	0.7756	0.8554
Number of iterations	1,190	574	341	344

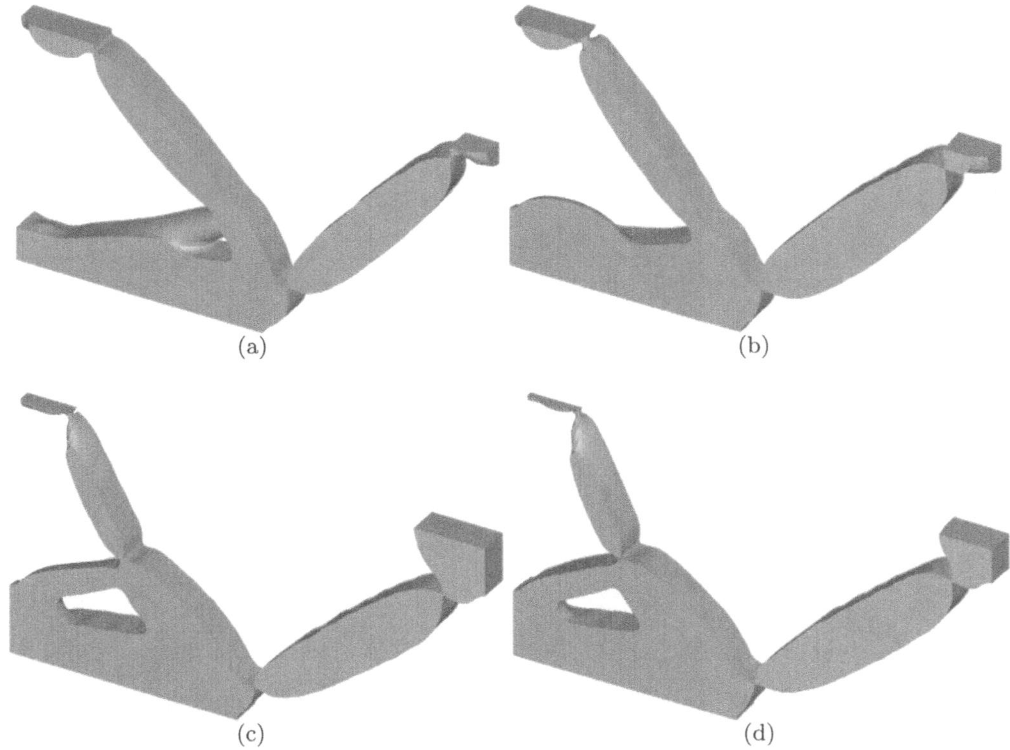

FIGURE 6.9
Resulting topologies of 3D force inverter design case under different input spring stiffnesses: (a) k_{in}=0.1, (b) k_{in}=0.01, (c) k_{in}=0.001, and (d) k_{in}=0.0001.

very different final 2D topological configurations. It is not clear if there is any trend in features of the mechanism. This variability in structure combined with the thin features connected to the input actuator will cause some difficulty in the fabrication of the mechanism structures.

The output displacement and convergence of the 3D force inverter design case under the various output spring stiffnesses are outlined in Table 6.6. As summarized in Table 6.6, when $k_{\text{out}} = 0.1$, the convergence is slowest and the number of iterations is the largest (1149), and with the decrease in k_{out}, the output displacement rises. The resulting topological configurations of the 3D force inverter design case when varying the output spring stiffnesses are shown in Figure 6.11, where very distinct topological configurations are observed.

TABLE 6.5
Output displacement and convergence of 2D force inverter design case under different output spring stiffnesses

Output spring stiffness k_{out}	**0.1**	**0.01**	**0.001**	**0.0001**
Output displacement	0.0992	0.5370	1.4566	2.1019
Number of iterations	480	310	378	578

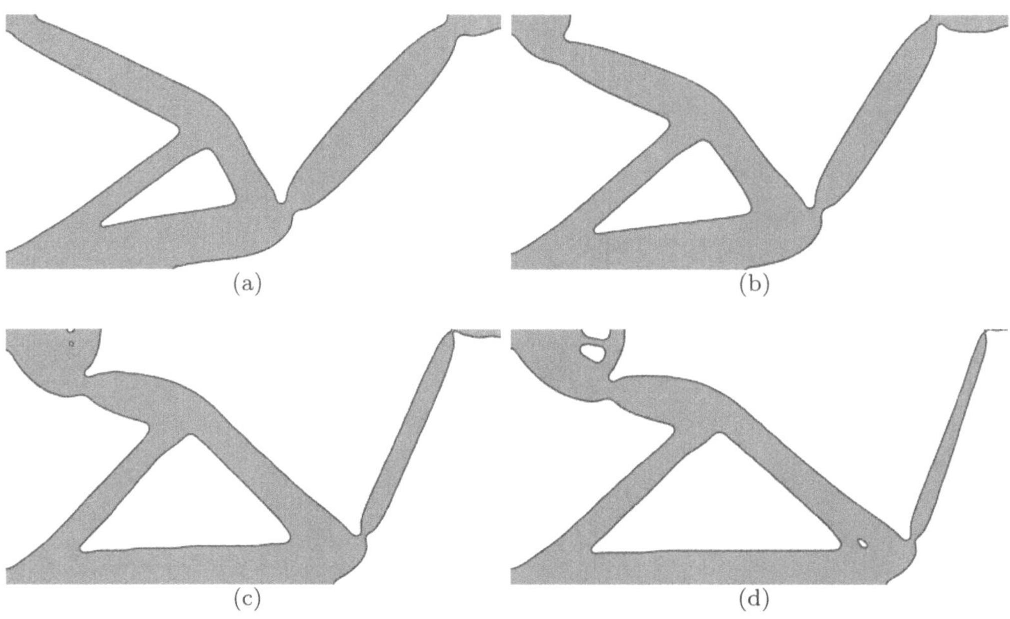

FIGURE 6.10
Resulting topologies of 2D force inverter design case under different output spring stiffnesses: (a) k_{out}=0.1, (b) k_{out}=0.01, (c) k_{out}=0.001, and (d) k_{out}=0.0001.

TABLE 6.6
Output displacement and convergence of 3D force inverter design case under different output spring stiffnesses

Output spring stiffness k_{out}	**0.1**	**0.01**	**0.001**	**0.0001**
Output displacement	0.1625	0.5743	0.8757	0.9031
Number of iterations	1,149	554	615	328

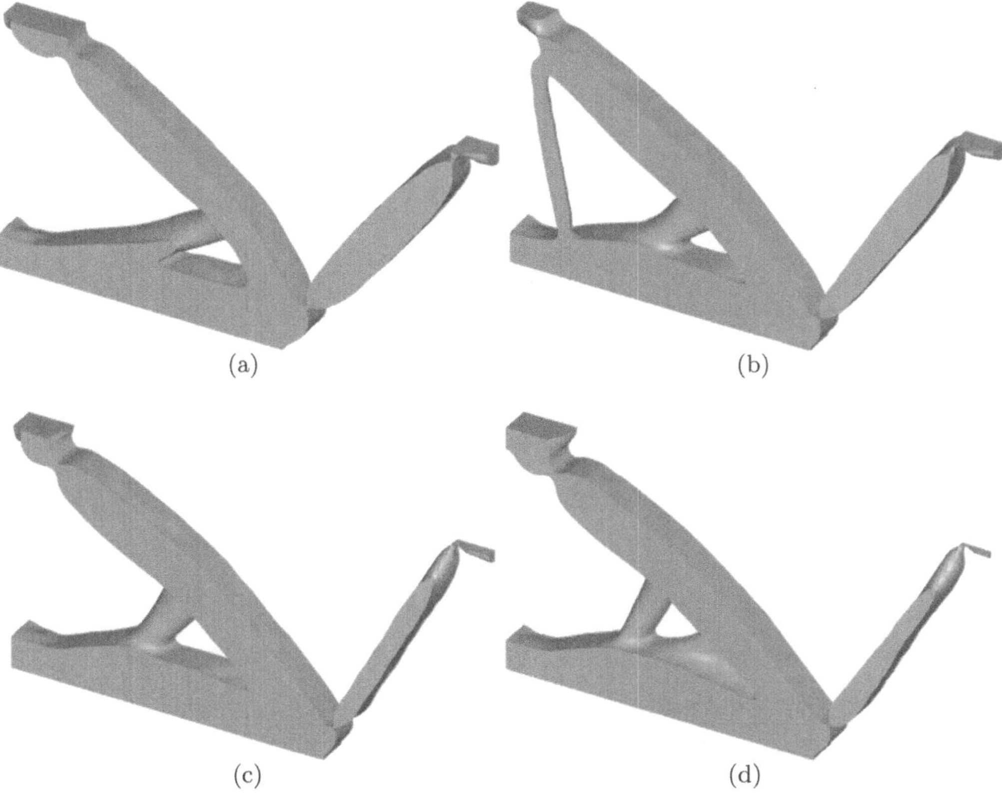

FIGURE 6.11
Resulting topologies of 3D force inverter design case under different output spring stiffnesses: (a) k_{out}=0.1, (b) k_{out}=0.01, (c) k_{out}=0.001, and (d) k_{out}=0.0001.

6.3 Summary

This chapter investigated Smooth-Edged Material Distribution for Optimizing Topology (SEMDOT) on the compliant mechanism design problem using the MMA optimizer. The move limit in MMA has only a slight effect on the 2D case but it has significant effect on the 3D case. The authors suggest that a move limit should be chosen that maintains a balance between performance and convergence. The authors recommend a move limit of 0.25 for the compliant mechanism design problem using SEMDOT. When decreasing either the input or output spring stiffnesses, the output displacement increases. However, when varying the input or output spring stiffnesses, the resulting topological configurations are quite distinct.

7

Heat Conduction Problems

This chapter presents the application of Smooth-Edged Material Distribution for Optimizing Topology (SEMDOT) to steady-state and transient heat conduction problems. In the steady-state heat conduction problem, the influences of target volume fractions, filter radius, and evolution rate on the Heaviside smooth function for thermal compliance, the total number of iterations, and final topological configuration are investigated via both 2D and 3D cases. In the transient heat conduction problem, the effects of varying the heat load working time on thermal compliance, the total number of iterations, and final topological configuration are studied through a representative 2D transient case.

7.1 Steady-State Heat Conduction Problems

7.1.1 Optimization Problem

Heat conduction is the process where heat transfers from hot to cold regions in a system. Heat conduction can occur anywhere and at any time in nature. The optimal heat management in a system is of great importance for real-world applications, such as heat diffusers, cooling fins, molding dies, and high-conductivity channels of electronic components [4,6,11,55]. Topology optimization has become an important tool used to explore the optimal structural configuration for a heating system to achieve specific goals [121,206]. The steady-state heat conduction case is generally used to validate the effectiveness of a topology optimization algorithm. The goal is to minimize the thermal compliance, which is equivalent to minimizing the average temperature. The optimization problem is mathematically described as:

$$
\begin{aligned}
&\min : C(X_e) = \varphi = \mathbf{P}^{\mathbf{T}}\mathbf{U} \\
&\text{subject to} : \frac{\sum_{e=1}^{M} X_e V_e}{\sum_{e=1}^{M} V_e} - V^* \leq 0, \\
&0 < \rho_{\min} \leq X_e \leq 1; \ e = 1, 2, \cdots, M,
\end{aligned}
\tag{7.1}
$$

where $C(X_e)$ is the objective function; X_e is the elemental volume fraction; φ is the thermal compliance; $\mathbf{U}$ is the nodal temperature vector; $\mathbf{P}$ is the global thermal load vector which has the form of $\mathbf{P} = \tilde{\mathbf{K}}\mathbf{U}$, where $\tilde{\mathbf{K}}$ is the thermal conductivity matrix; M is the total number of elements; V_e is the volume of the eth element; V^* is the target volume; and $\rho_{\min}$ is the minimum elemental volume fraction.

DOI: 10.1201/9781032634449-7

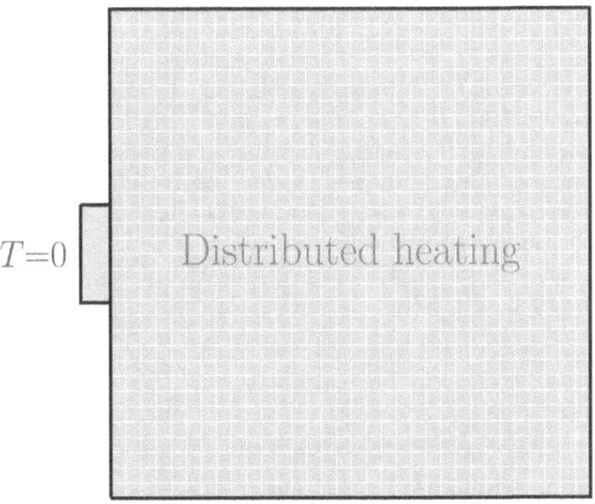

FIGURE 7.1
2D heat conduction problem.

Sensitivities of the solid and void elements for the heat conduction problem are [207]:

$$\frac{\partial C(X_e)}{\partial X_e} = \begin{cases} \dfrac{C(X_e=1)-C(X_e=0)}{1} \approx -\mathbf{U}_e^{\mathbf{T}}\tilde{\mathbf{K}}_e^{1}\mathbf{U}_e & \text{if } X_e=1, \\ \dfrac{C(X_e=0)-C(X_e=1)}{-1} \approx -\rho_{\min}\mathbf{U}_e^{\mathbf{T}}\tilde{\mathbf{K}}_e^{1}\mathbf{U}_e & \text{if } X_e=\rho_{\min}, \end{cases} \tag{7.2}$$

where $\mathbf{U}_e$ is the temperature of the eth element and $\tilde{\mathbf{K}}_e^1$ is the thermal conductivity matrix of the solid element.

Sensitivities of boundary elements are calculated as:

$$\begin{aligned} \frac{\partial C(X_e)}{\partial X_e} \approx& (1-X_e)\left.\frac{\partial C(X_e)}{\partial X_e}\right|_{X_e=\rho_{\min}} + X_e\left.\frac{\partial C(X_e)}{\partial X_e}\right|_{X_e=1} \\ =& -[(1-X_e)\rho_{\min} + X_e]\mathbf{U}_e^{\mathbf{T}}\tilde{\mathbf{K}}_e^{1}\mathbf{U}_e. \end{aligned} \tag{7.3}$$

7.1.2 Numerical Examples

7.1.2.1 Effects of Volume Fraction

In this section, the effects of varying the volume fraction V^* on performance and convergence are investigated. The 2D optimization case is illustrated in Figure 7.1, where the square plate is evenly heated and the center of the left edge is a heat sink, meaning that the temperature is set to zero. Two material phases with isotropic conductivities of 1 and 0.001 are taken into account. A mesh of 200×200 is used, the target volume V^* is set to 0.5, and the filter radius $r_{\min}$ is set to 3.5. In addition, the move limit in Method of Moving Asymptote (MMA) is set to 1. To stabilize the optimization process, a scaling factor of 0.1 is employed in the MMA optimizer. The convergence process and optimized topology of the 2D heat conduction case are shown in Figure 7.2. The optimization process converges at the thermal compliance of 3195 after 721 iterations.

The thermal compliance and convergence of the 2D heat conduction case under the range of target volume fractions (V^*=0.15, 0.2, 0.3, 0.4, 0.6, and 0.8) are summarized in

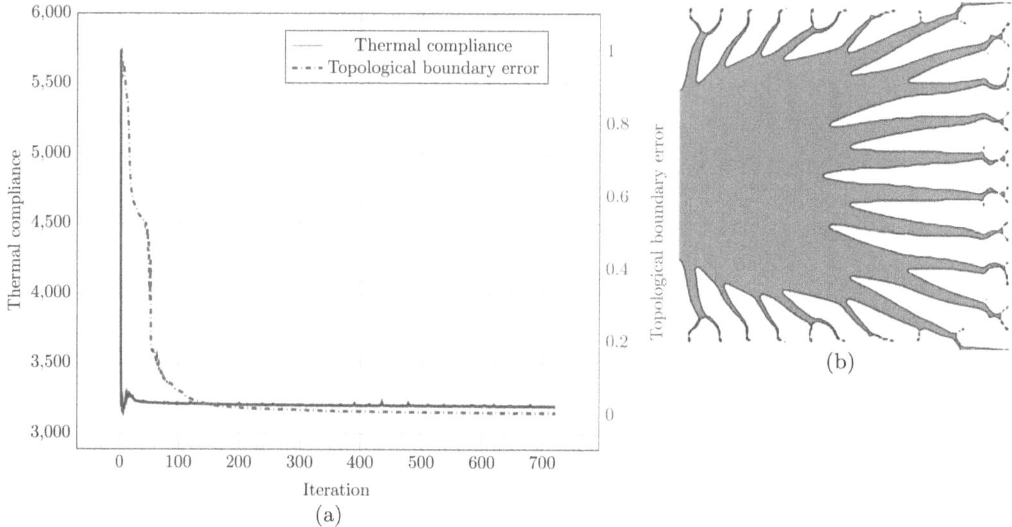

FIGURE 7.2
2D heat conduction case: (a) convergence process and (b) optimized topology.

Table 7.1. The data shows the self-evident relationship that when the target volume fraction increases the thermal compliance decreases. There is no clear relation between the number of iterations and target volume fraction. The topological configurations under the range of target volume fractions are shown in Figure 7.3 where tree-like structures are found.

The 3D optimization case is illustrated in Figure 7.4 where the cubic design domain is homogeneously heated, and the center of the bottom surface is a heat sink, namely the temperature is set to zero. To stabilize the optimization process, a scaling factor of 0.001 is used in the MMA optimizer. The move limit of 1 in MMA is used. A mesh of 60×60×60 and $r_{\min} = 3.5$ is used. When considering $V^* = 0.5$, the convergence process and optimized topology structure are shown in Figure 7.5. Figure 7.5(a) shows that the optimization process converges at the thermal compliance of 343,200 after 382 iterations. The resulting topology is shown in Figure 7.5(b).

The thermal compliance and total number of iterations of the 3D heat conduction case under a range of target volume fractions (V^*=0.15, 0.2, 0.3, 0.4, 0.6, and 0.8) are summarized in Table 7.2. Table 7.2 confirms the self-evident relationship that when there is an increase in target volume fraction, then there is a decrease in the thermal compliance. The 3D heat conduction case generally requires a long convergence process since numerous structural features are formed. The resulting topologies under the range of target volume fractions are shown in Figure 7.6 where, again, tree-like structures are observed.

TABLE 7.1
Thermal compliance and convergence of 2D heat conduction case under different target volume fractions

Target volume fraction V^*	**0.15**	**0.2**	**0.3**	**0.4**	**0.6**	**0.8**
Thermal compliance	7,317.72	5,691.29	4,110.23	3,489.21	3,042.04	2,904.47
Number of iterations	641	575	819	615	646	663

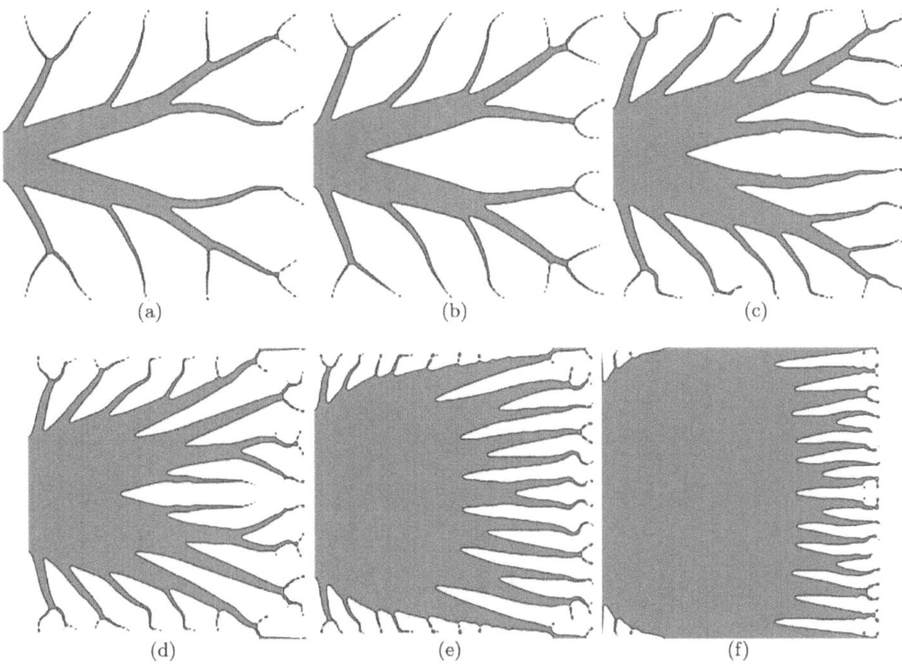

FIGURE 7.3
Resulting topologies of 2D heat conduction case under a range of target volume fractions: (a) $V^* = 0.15$, (b) $V^* = 0.2$, (c) $V^* = 0.3$, (d) $V^* = 0.4$, (e) $V^* = 0.6$, and (f) $V^* = 0.8$.

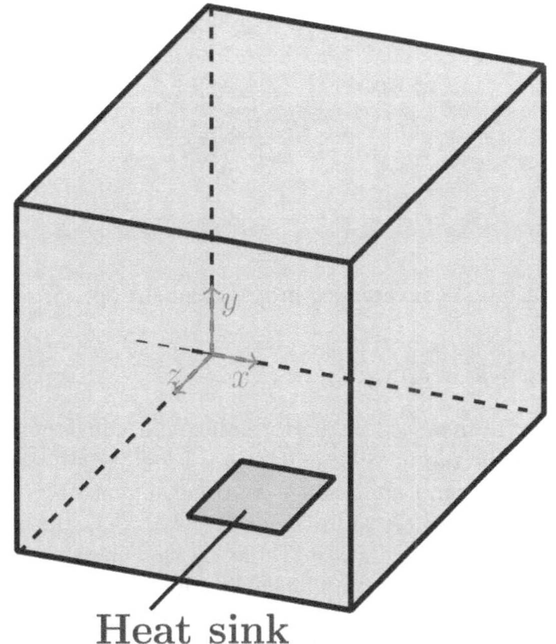

FIGURE 7.4
3D heat conduction problem.

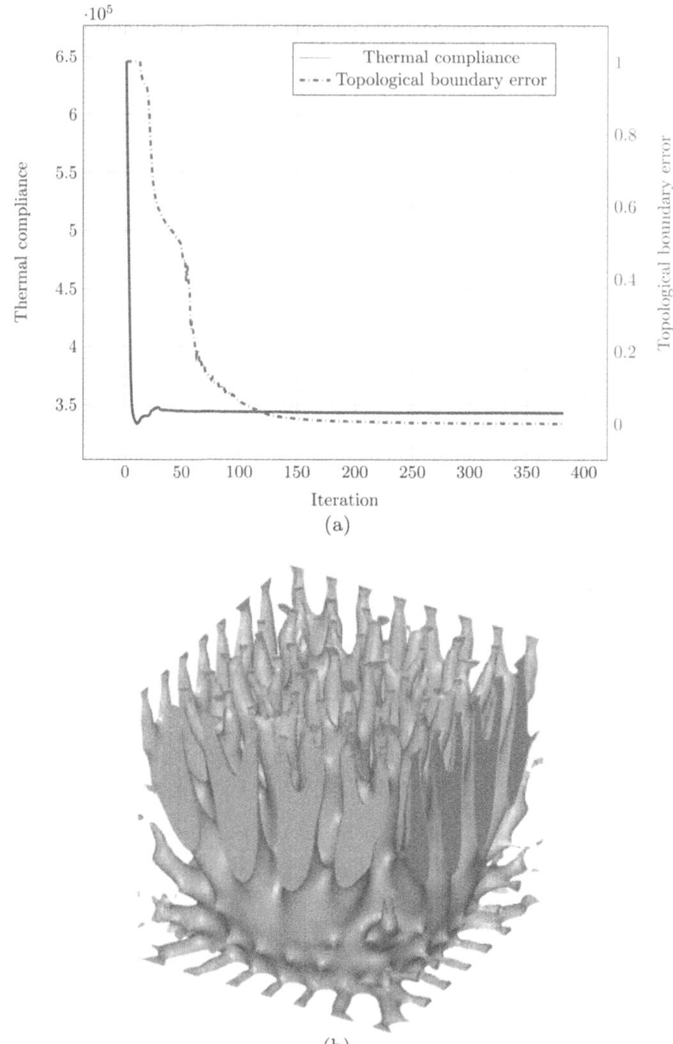

FIGURE 7.5
3D heat conduction case: (a) convergence process and (b) optimized topology.

7.1.2.2 Effects of Filter Radius

The effects of the filter radius $r_{\min}$ on performance and convergence are studied here. In the 2D case, other than the target volume fraction, $V^* = 0.3$, the parameters are the same as Section 7.1.2.1. The thermal compliance and convergence of the 2D heat conduction case results when varying the filter radii ($r_{\min}$=1.5, 2.5, 3, 4, 5, and 6) are summarized in Table 7.3. It can be seen from the data in Table 7.3 that increases in the filter radius also see general increases in the thermal compliance. In addition, there is no obvious relationship between the filter radii and number of iterations. The resulting topologies under a range of filter radii are shown in Figure 7.7. Generally, when the filter radius increases, then the topological configuration becomes simpler.

TABLE 7.2
Thermal compliance and convergence of 3D heat conduction case under different target volume fractions

Target volume fraction V^*	**0.15**	**0.2**	**0.3**	**0.4**	**0.6**	**0.8**
Thermal compliance	589,512	480,590	395,451	360,656	334,274	325,223
Number of iterations	1,569	893	1,709	797	389	537

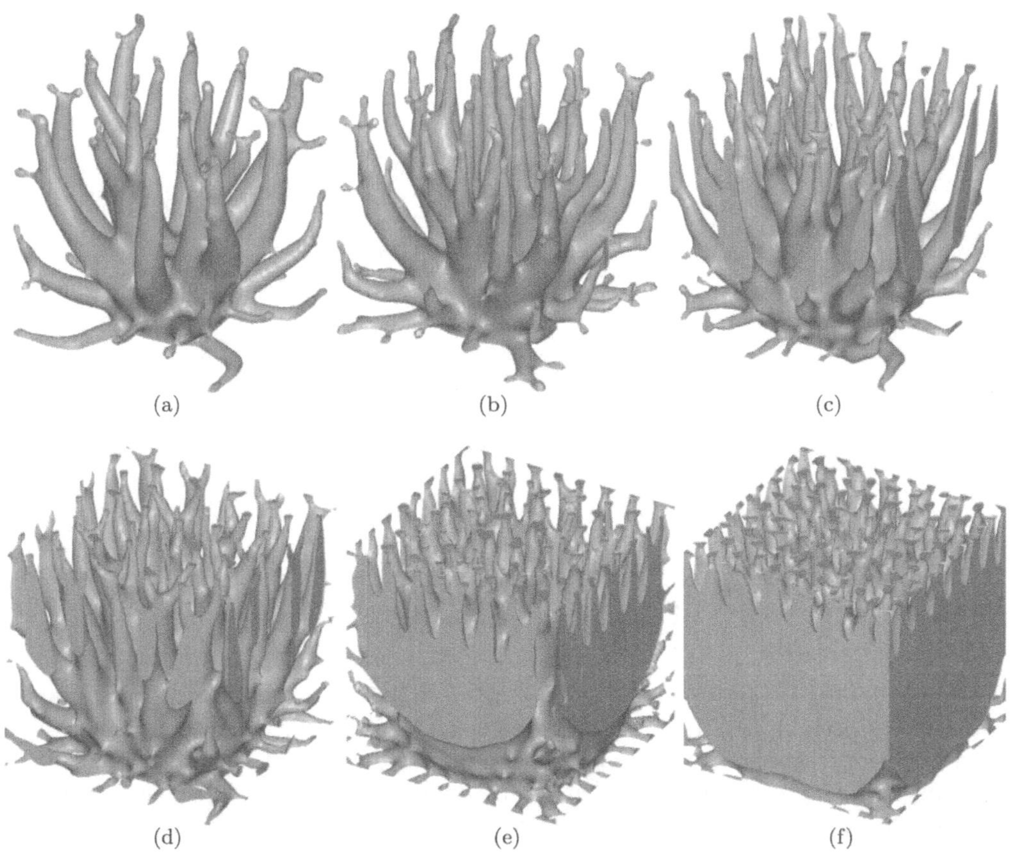

FIGURE 7.6
Resulting topologies of 3D heat conduction case under a range of target volume fractions: (a) $V^* = 0.15$, (b) $V^* = 0.2$, (c) $V^* = 0.3$, (d) $V^* = 0.4$, (e) $V^* = 0.6$, and (f) $V^* = 0.8$.

TABLE 7.3
Thermal compliance and convergence of 2D heat conduction case under different filter radii

Filter radius $r_{\min}$	**1.5**	**2.5**	**3**	**4**	**5**	**6**
Thermal compliance	4,076.62	4,098.77	4,106.12	4,095.99	4,188.70	4,225.32
Number of iterations	2,319	428	556	915	1,469	1,742

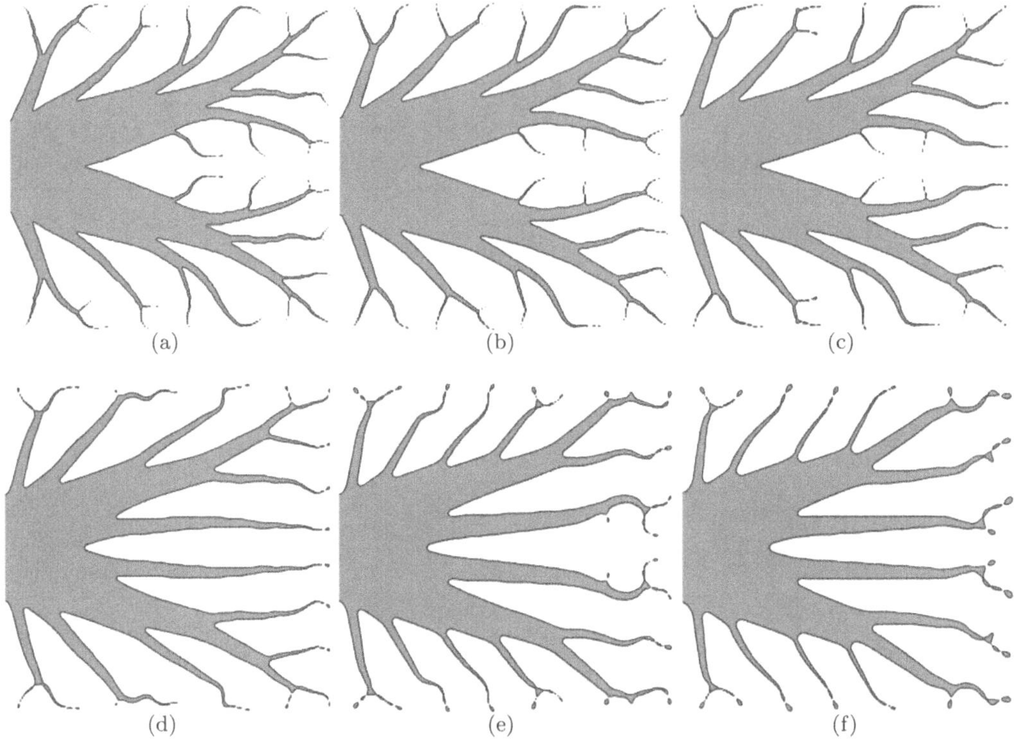

FIGURE 7.7
Resulting topologies of 2D heat conduction case under a range of filter radii: (a) $r_{\min} = 1.5$, (b) $r_{\min} = 2.5$, (c) $r_{\min} = 3$, (d) $r_{\min} = 4$, (e) $r_{\min} = 5$, and (f) $r_{\min} = 6$.

In the 3D case, other than the target volume fraction, $V^* = 0.2$, the parameters are the same as Section 7.1.2.1. The thermal compliance and number of iterations of the 3D heat conduction case under a range of filter radii ($r_{\min}$=1.5, 2.5, 3, 4, 5, and 6) are outlined in Table 7.4. The results indicate that when the filter radius increases, then the thermal compliance rises. There is no apparent relationship between the number of iterations and filter radius. The resulting topologies when varying the filter radii are shown in Figure 7.8, where the increase in the filter radius correlates to simpler topological configurations.

7.1.2.3 Effects of Evolution Rate in Heaviside Smooth Function

Effects of the evolution rate, Λ, on thermal compliance and convergence are discussed here. In the 2D case, other than the target volume fraction, $V^* = 0.3$, and filter radii, $r_{\min} = 4$, the parameters are the same as Section 7.1.2.1. The effects on thermal compliance and convergence from the 2D heat conduction case when varying the evolution rates ($\Lambda = 0.1$,

TABLE 7.4
Thermal compliance and convergence of 3D heat conduction case under a range of filter radii

Filter radius $r_{\min}$	**1.5**	**2.5**	**3**	**4**	**5**	**6**
Thermal compliance	454,956	466,962	469,937	497,660	521,248	537,050
Number of iterations	119	384	1,300	879	637	640

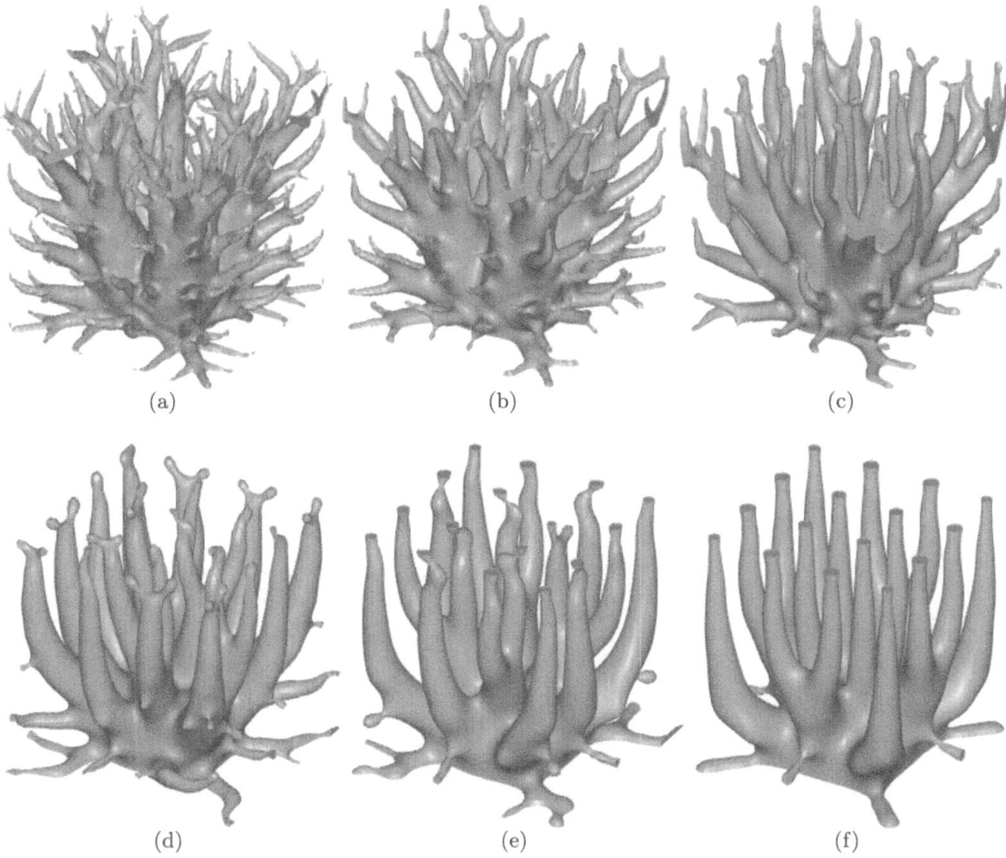

FIGURE 7.8
Resulting topologies of 3D heat conduction case under a range of filter radii: (a) $r_{\text{min}} = 1.5$, (b) $r_{\text{min}} = 2.5$, (c) $r_{\text{min}} = 3$, (d) $r_{\text{min}} = 4$, (e) $r_{\text{min}} = 5$, and (f) $r_{\text{min}} = 6$.

0.25, 0.75, 1, 1.5, and 2) are summarized in Table 7.5. The results indicate that an increase in the evolution rate generally leads to an increase in the thermal compliance. In addition, a low value of the evolution rate leads to a long convergence process. By contrast, a high value of the evolution rate can accelerate the optimization process; however, there is no guarantee that a faster optimization provides the same thermal compliance and convergence as a slower optimization. The optimized topologies of the 2D heat conduction case under a range of evolution rates are shown in Figure 7.9. It can be seen from Figure 7.9 that the evolution rate can significantly affect the final optimized topological configuration.

In the 3D case, other than the target volume fraction, $V^* = 0.2$, and the filter radii, $r_{\text{min}} = 4$, the parameters are the same as Section 7.1.2.1. The effects on thermal compliance

TABLE 7.5
Thermal compliance and convergence of 2D heat conduction case under a range of evolution rates

Evolution rate Λ	**0.1**	**0.25**	**0.75**	**1**	**1.5**	**2**
Thermal compliance	4,027.79	4,109.46	4,149.74	4,132.80	4,206.01	4,253.92
Number of iterations	4,092	1,694	753	907	715	337

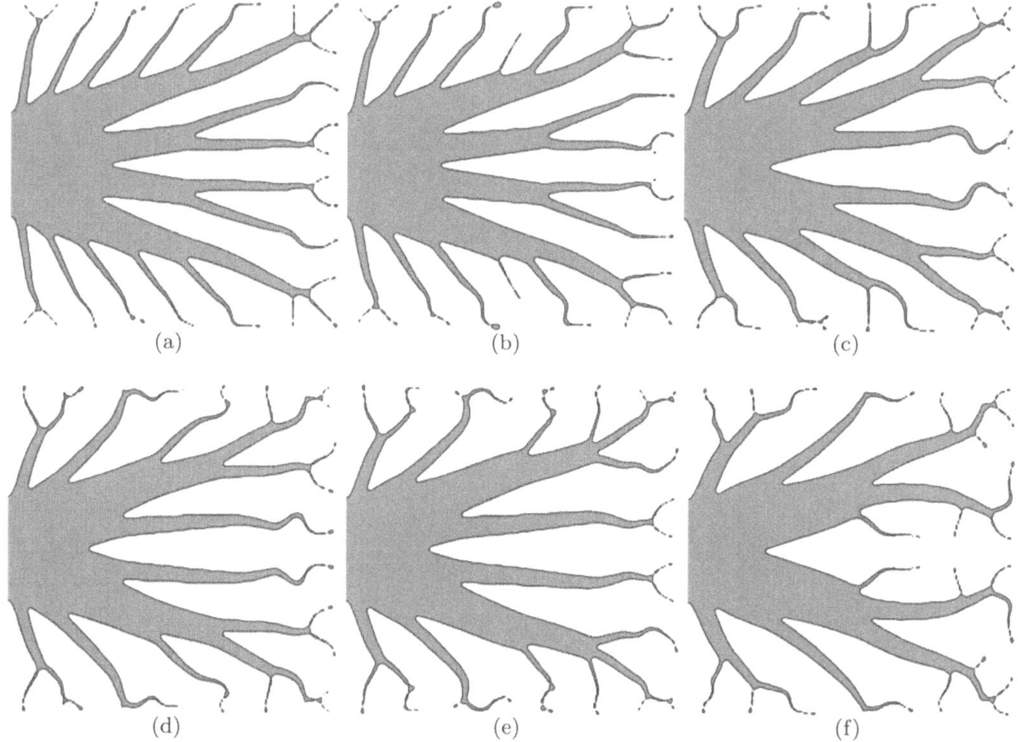

FIGURE 7.9
Resulting topologies of 2D heat conduction case under a range of evolution rates: (a) $\Lambda = 0.1$, (b) $\Lambda = 0.25$, (c) $\Lambda = 0.75$, (d) $\Lambda = 1$, (e) $\Lambda = 1.5$, and (f) $\Lambda = 2$.

and the number of iterations from the 3D heat conduction case when varying evolution rates ($\Lambda = 0.1$, 0.25, 0.75, 1, 1.5, and 2) are listed in Table 7.6. The results show that a low value of the evolution rate generally leads to a long convergence process. Although as has been previously suggested in the 2D case, increasing the evolution rate can speed up the optimization process; however, the thermal performance may be sacrificed in the process. The resulting topologies of the 3D heat conduction case under a range of evolution rates are shown in Figure 7.10. It can be seen from Figure 7.10 that each evolution rate can lead to an obviously distinct topological configuration.

7.1.2.4 Remarks on the Discontinuous Boundary Issue and Its Solution

When Smooth-Edged Material Distribution for Optimizing Topology (SEMDOT) is used to solve heat conduction problems, the obvious limitation is the structural disconnection

TABLE 7.6
Thermal compliance and convergence of 3D heat conduction case under a range of evolution rates

Evolution rate Λ	**0.1**	**0.25**	**0.75**	**1**	**1.5**	**2**
Thermal compliance	491,721	491,870	495,323	497,703	496,762	496,746
Number of iterations	2,068	2,260	799	779	859	827

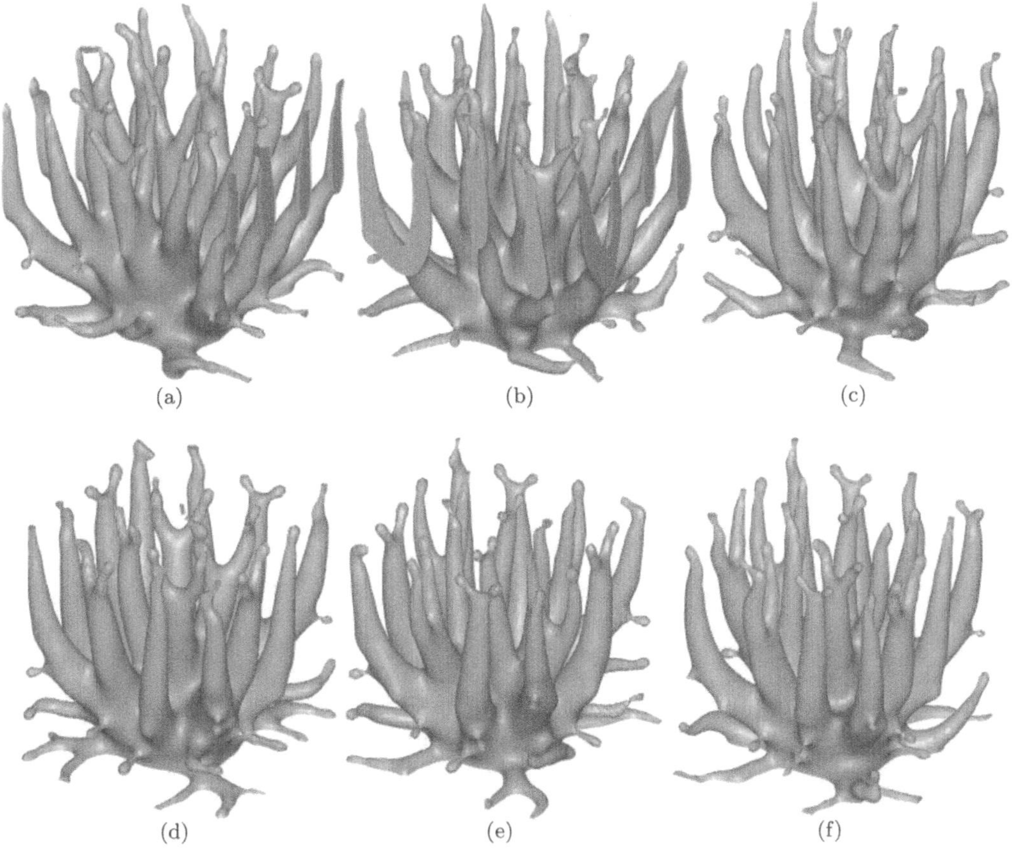

FIGURE 7.10
Resulting topologies of 3D heat conduction case under a range of evolution rates: (a) $\Lambda = 0.1$, (b) $\Lambda = 0.25$, (c) $\Lambda = 0.75$, (d) $\Lambda = 1$, (e) $\Lambda = 1.5$, and (f) $\Lambda = 2$.

or breakage. Structural disconnection occurs when there is no connection between a load and the boundaries conditions of the system. The discontinuous boundary issue in the 2D case is more severe than in the 3D case. The discontinuous boundary issue cannot be easily solved by increasing the number of elements or grid points. To overcome this issue, the length scale control approach proposed by Zhang et al. [213] is adopted to explicitly control the minimum feature size. In this method, the minimum length scale can be accurately measured based on the structural skeleton that is identified by means of the skeletonization image processing technique. The minimum length scale constraint is defined as :

$$g^*(X_e) = \sum_{j \in I_{\min}} (X_j - 1)^2 \leq 0, \tag{7.4}$$

where

$$I_{\min} = \left\{ j \,\middle|\, j \in \{1, \cdots, M\}, \Omega_j \in \bigcup_{\mathbf{X} \in \mathcal{SS}(\Omega)} \mathfrak{B}(\mathbf{X}, \underline{d}) \right\}, \tag{7.5}$$

and X_j is the elemental volume fraction of the jth element, $\mathbf{X}$ is the vector of X_j, Ω is a closed domain, Ω_j is the domain occupied by the jth element, $\mathcal{SS}(\Omega)$ is the structural

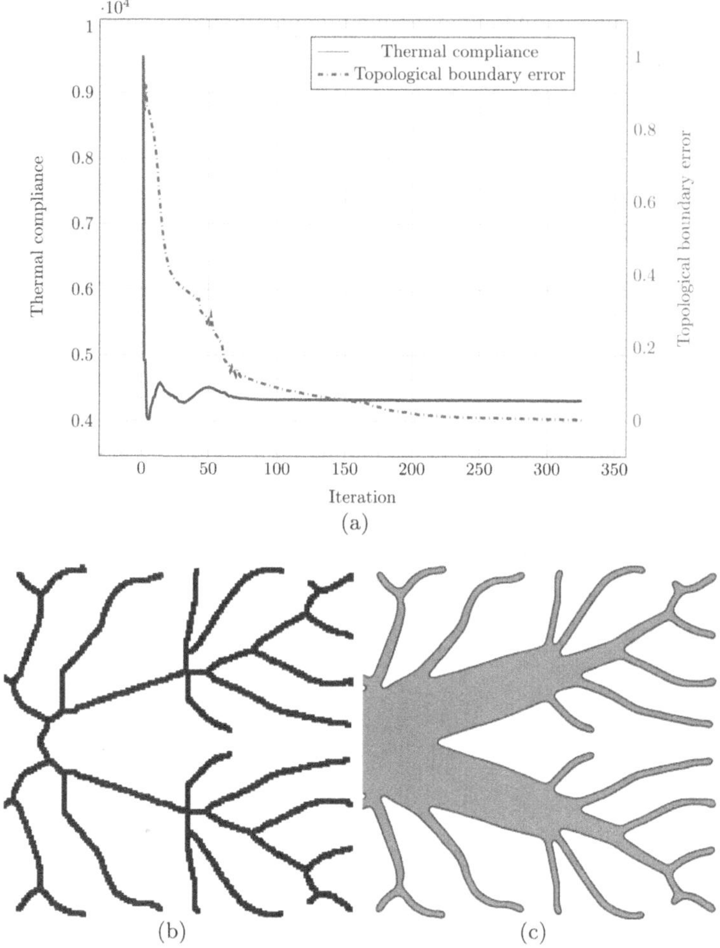

FIGURE 7.11
(a) Convergence process, (b) skeleton, and (c) topology obtained by length scale control approach for 2D steady-state heat conduction problem.

skeleton, $\underline{d}$ is the lower bound of the length scale, and $\mathfrak{B}(\mathbf{X}, \underline{d})$ is a closed ball centered at $\mathbf{X}$ with a diameter of $\underline{d}$.

A 2D heat conduction case with a mesh of 200×200, target volume fraction, $V^* = 0.3$, and lower bound of the length scale, $\underline{d} = 4$ is taken as an example. The convergence process, skeleton, and optimized topology obtained by the length scale control approach for the 2D steady-state heat conduction problem are shown in Figure 7.11. Figure 7.11(a) shows that the optimization process converges at the thermal compliance of 4,314 after 326 iterations. The final skeleton and topology configuration are shown in Figure 7.11(b) and (c), respectively. More importantly, there is no discontinuous boundary issue in the final topology presented in Figure 7.11(c).

Other than the approach proposed by Zhang et al. [213], similar feature size control methods can be used to solve the discontinuous boundary issue when using SEMDOT and steady-state heat conduction problems.

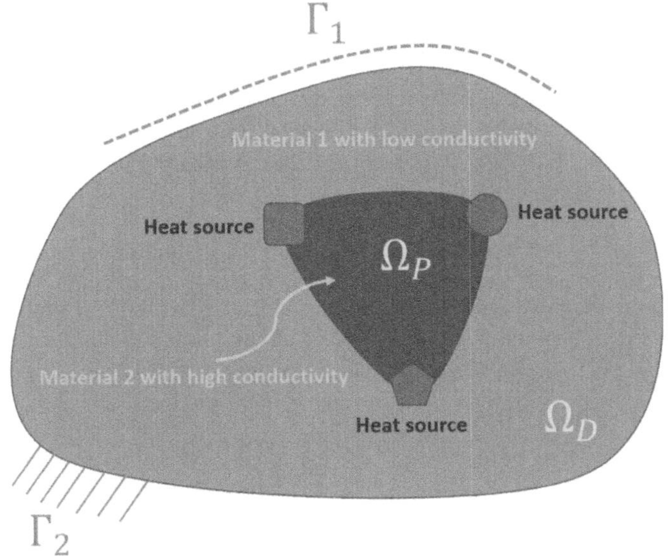

FIGURE 7.12
Transient heat conduction with two-phase materials.

7.2 Transient Heat Conduction Problems

7.2.1 Optimization Problem

Up to this point of the chapter, we have been investigating steady-state heat conduction problems. However, the majority of practical heat conduction problems are transient in nature, and the temperature varies with time in such applications [210, 222]. Therefore, the way of implementing topology optimization for minimizing the maximum temperature of transient heat conduction structures using SEMDOT is presented here. The contents in this section are mainly based on the work in [108, 124, 191, 192, 222]. The transient heat conduction problem with two-phase materials is illustrated in Figure 7.12, where the domain that is heated by the discrete (or uniform) heat sources are bounded by the solid line. In Figure 7.12, there are two sorts of conditions: the constant temperature and adiabatic boundary conditions. The constant temperature condition is represented by the dashed line, and the adiabatic boundary restricts heat flux out of the domain. The material with low conductivity and high heat capacity is filled across the whole design domain. To prevent an excessive temperature, the material with high thermal conductivity is inserted such that the thermal diffusion efficiency can be improved. The transient-state conductive heat transfer within the domain is mathematically described by the governing equations:

$$\begin{aligned}
&\rho H_s \frac{\partial T}{\partial t} - \nabla \cdot (k \nabla T) - \rho Q = 0 \quad \text{on } \Omega_D, \\
&T = \bar{T} \quad \text{on } \Gamma_1, \\
&-k \frac{\partial T}{\partial n} = q \quad \text{on } \Gamma_2, \\
&T|_{t=0} = T_0 \quad \text{on } \Omega_D,
\end{aligned} \tag{7.6}$$

where T is the structural temperature field, ρ is the material density, H_s is the specific heat of the material, t is the time variable of the transient process, k is the heat conductivity, Q is the heat energy generated per unit mass inside the object, n is the unit outward normal vector of boundary, $\bar{T} = \bar{T}(\Gamma, t)$ is a prescribed temperature on the boundary Γ_1, $q = q(\Gamma, t)$ is a prescribed heat flux on the boundary Γ_2, and T_0 is the initial ambient temperature.

Based on finite element theory, the control equation for the transient heat conduction problem (Equation 7.6) can be approximated as:

$$\mathbf{H}\dot{\mathbf{T}}_\mathbf{n} + \mathbf{K}\mathbf{T}_\mathbf{n} = \mathbf{P}_\mathbf{T} \text{ with } \mathbf{T}_\mathbf{n}|_{t=0} = \mathbf{T}_\mathbf{n}^\mathbf{0}, \tag{7.7}$$

where $\mathbf{H}$ is the heat capacity matrix, $\mathbf{K}$ is the thermal conductivity matrix, $\mathbf{T}_\mathbf{n}$ is the node temperature vector, $\dot{\mathbf{T}}_\mathbf{n}$ is the derivative vector of the node temperature to time, $\mathbf{P}_\mathbf{T}$ is the imposed heat load vector, and $\mathbf{T}_\mathbf{n}^\mathbf{0}$ is the temperature vector for the initial state.

A finite difference technique is used for the time approximation to solve Equation (7.7), which is given by:

$$\dot{\mathbf{T}}_\mathbf{n} = \frac{{\mathbf{T}_\mathbf{n}}^{i+1} - {\mathbf{T}_\mathbf{n}}^{i}}{\Delta t} + O(\Delta t), \tag{7.8}$$

where $\Delta t = t_{i+1} - t_i$ and ${\mathbf{T}_\mathbf{n}}^{i+1}$ and ${\mathbf{T}_\mathbf{n}}^{i}$ are the temperatures at the $(i+1)$th and ith levels, respectively.

A linear interpolation scheme and a parameter θ are introduced for the temperature field and load vector, which has the form:

$${\mathbf{T}_\mathbf{n}}^{i+\theta} = \theta {\mathbf{T}_\mathbf{n}}^{i+1} + (1-\theta){\mathbf{T}_\mathbf{n}}^{i}, {\mathbf{P}_\mathbf{T}}^{i+\theta} = \theta {\mathbf{P}_\mathbf{T}}^{i+1} + (1-\theta){\mathbf{P}_\mathbf{T}}^{i}. \tag{7.9}$$

Substituting Equations (7.8) and (7.9) into Equation (7.7), the system equation can be calculated by:

$$(\mathbf{H} + \theta\Delta t\mathbf{K}){\mathbf{T}_\mathbf{n}}^{i+1} = [\mathbf{H} - (1-\theta)\Delta t\mathbf{K}]{\mathbf{T}_\mathbf{n}}^{i} + \Delta t[\theta {\mathbf{P}_\mathbf{T}}^{i+1} + (1-\theta){\mathbf{P}_\mathbf{T}}^{i}]. \tag{7.10}$$

Once the initial temperature field is given, the temperature field at an arbitrary time can be calculated by solving Equation (7.10) in successive steps.

The thermal conductivity and heat capacity of the eth element are expressed by:

$$\begin{aligned} K_e &= (1 - X_e)K_1 + X_e K_2, \\ H_e &= (1 - X_e)H_1 + X_e H_2, \end{aligned} \tag{7.11}$$

where K_e is the thermal conductivity of the eth element, H_e is the heat capacity of the eth element, K_1 and H_1 are the thermal conductivity and heat capacity of Material-1, respectively, and K_2 and H_2 are the thermal conductivity and heat capacity of Material-2, respectively.

The constraint on the volume of Material-2 is expressed by:

$$\phi\Omega_D = \sum_{e=1}^{M} X_e A_e, \tag{7.12}$$

where ϕ is the volume fraction of the high-conductivity material and A_e is the area of the eth element.

The thermal compliance can be defined by integrating the transient thermal compliance over the time interval:

$$C(t) = \int_0^{t_T} \mathbf{T}_\mathbf{n}^\mathbf{T}(t)\mathbf{K}\mathbf{T}_\mathbf{n}(t)\mathrm{d}t, \tag{7.13}$$

where t_T is the whole working time and $\mathbf{T}_\mathbf{n}^\mathbf{T}$ is the transpose of the vector $\mathbf{T}_\mathbf{n}$.

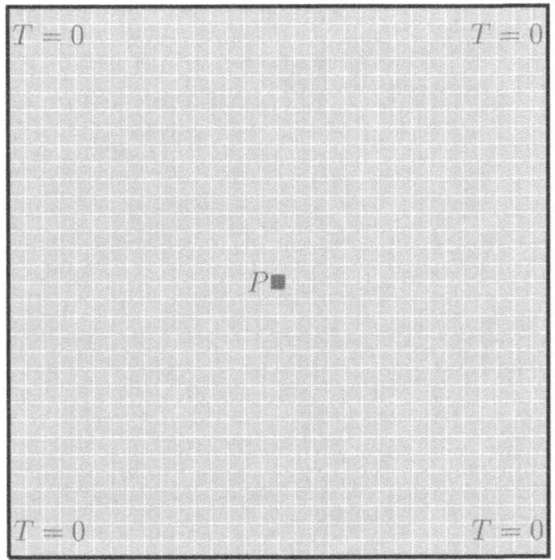

FIGURE 7.13
Transient heat conduction problem.

Taking the thermal compliance as the objective, the optimization problem is therefore expressed by:

$$\begin{aligned} \min : C(t) &= \int_0^{t_T} \mathbf{T_n^T}(t)\mathbf{K}\mathbf{T_n}(t)\mathrm{d}t \\ \text{subject to} : \mathbf{H}\dot{\mathbf{T}}_\mathbf{n} &+ \mathbf{K}\mathbf{T_n} = \mathbf{P_T}, \\ \frac{1}{A_t}\sum_{e=1}^{M} X_e A_e &= \phi, \end{aligned} \tag{7.14}$$

where $C(t)$ is the transient heat compliance and A_t is the total volume of the design domain.

There are a range of ways to execute sensitivity analysis for transient heat conduction cases. More details can be found in [108, 124, 191, 192, 222].

7.2.2 Numerical Examples

The design domain of the transient heat conduction problem is shown in Figure 7.13, where the dimension of the square structure with 0.1×0.1 and the heat flux of 0.1 are adopted. In addition, the temperatures of the four corners are fixed to 0 ($T = 0$). The material with low thermal conductivity (Material-1) is used across the design domain, and to improve the cooling efficiency, the material with a thickness of 0.001 (Material-2) is covered on the surface of the square structure. Specifically, the thermal conductivity of Material 2 is 100 times higher than that of Material 1. It is supposed that the coverage area is 20% of the whole design area. The properties of the two-phase materials are listed in Table 7.7. For all numerical examples, the move limit of 1 (that is, $mov = 1$) in the MMA optimizer is used.

TABLE 7.7
Properties of two-phase materials

	Thermal conductivity	**Heat capacity**
Material-1	0.1	5×10^5
Material-2	10	10^6

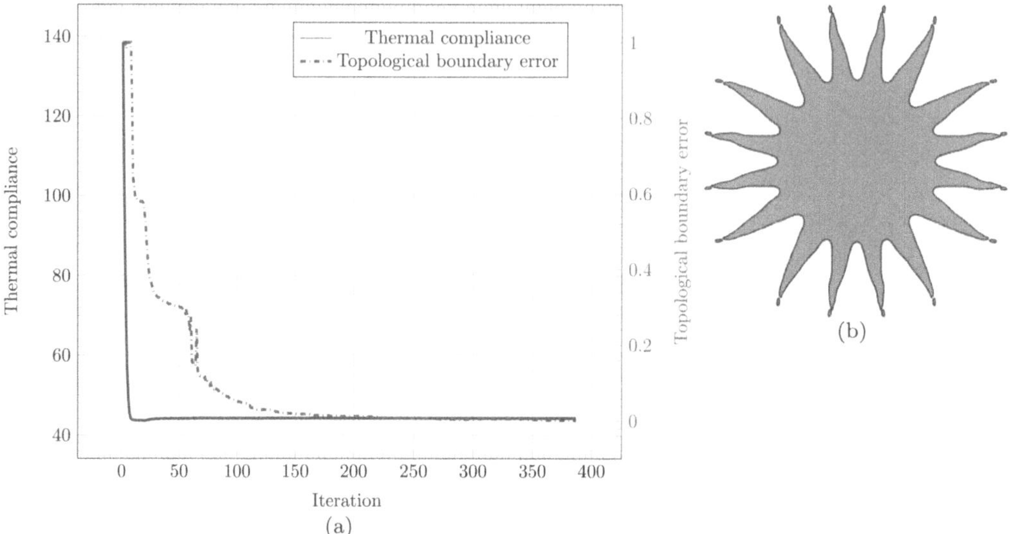

FIGURE 7.14
Transient heat conduction case subjected to working time of 1,000 s: (a) convergence process and (b) optimized topology

A mesh of 100×100, filter radius of 3 ($r_{\min} = 3$), and volume fraction of 0.3 for the high-conductivity material ($\phi = 0.3$) are adopted here. When the heat load working time is 1,000 s, the convergence process and optimized topology are shown in Figure 7.14. Figure 7.14(a) shows that the optimization process converges at the thermal compliance of 44 after 387 iterations. The optimized topology is shown in Figure 7.14(b).

The resulting thermal compliance values and number of iterations under a range of working times (500 s, 2,000 s, 3,000 s, 4,000 s, 5,000 s, and 6,000 s) are listed in Table 7.8. The results indicate that an increase in the working time leads to a rise in the thermal compliance. In addition, after a working time of 3,000 s and higher, the number of iterations remains steady at around 110 iterations. The resulting topologies of the transient heat conduction case under the range of working times are shown in Figure 7.15. After the working times of 3,000 s and higher, the topological configurations are only X-type. the topological configurations are relatively complex before a working time of 3,000 s. In addition, with the increase in the heat load working time, the stretching range of the heat conduction path gradually extends. It can be concluded that the topology optimization of the transient heat conduction structure exhibits a remarkable transient effect, meaning that the obtained topological configuration is closely related to the working time. Therefore, different working times of thermal loads may result in completely distinct topological configurations.

TABLE 7.8
Thermal compliance and convergence under a range of working times.

Working time	**500 s**	**2,000 s**	**3,000 s**	**4,000 s**	**5,000 s**	**6,000 s**
Thermal compliance	41.178	47.910	78.211	85.642	91.179	95.572
Number of iterations	446	436	115	114	114	115

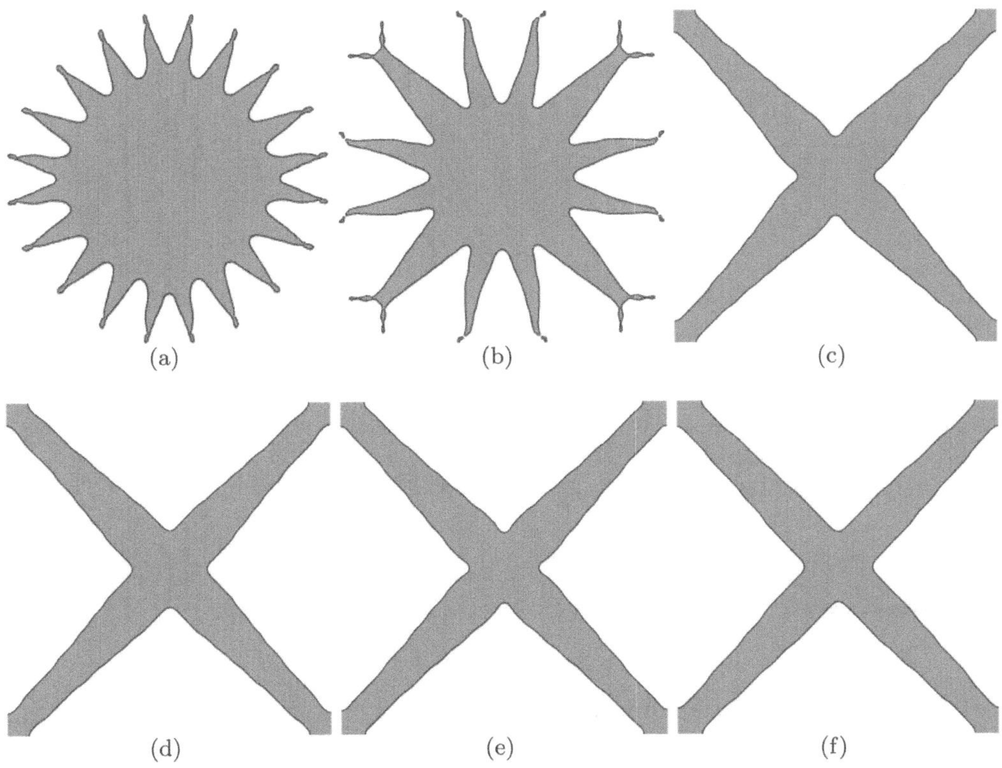

FIGURE 7.15
Resulting topologies of transient heat conduction case under a range of working times: (a) 500 s, (b) 2,000 s, (c) 3,000 s, (d) 4,000 s, (e) 5,000 s, and (f) 6,000 s.

7.3 Summary

The SEMDOT algorithm has been shown to successfully solve both steady-state and transient heat conduction problems. However, in the 2D steady-state heat conduction case, the discontinuous boundary issue was observed, and hence a length scale control approach was used to solve this issue. This chapter also showed that to reduce the complexity of the topological configuration, the filter radius can be increased. It was also shown that increasing the evolution rate in the Heaviside smooth function can accelerate the convergence process; however, this may sacrifice the thermal compliance of the system. In addition, topology optimization of the transient heat conduction structure exhibits a remarkable transient effect.

8

Dynamic Problems

This chapter conducts topology optimization of natural frequency maximization and dynamic compliance minimization problems using Smooth-Edged Material Distribution for Optimizing Topology (SEMDOT). Three representative cases (pin-pin, fixed-pin, and clamped beam cases) are used to demonstrate the capability of SEMDOT in solving natural frequency maximization problems. In addition, the effect of varying the frequency on final results of dynamic compliance minimization problems are investigated, and the effectiveness of SEMDOT in solving dynamic problems is demonstrated.

8.1 Natural Frequency Maximization Problems

8.1.1 Optimization Problem

Natural frequency optimization is very important for many engineering fields, in particular it is vital for the aeronautical and automotive industry sectors [82, 199]. The first recognized work to use topology optimization to solve dynamic problems was the investigation by Díaaz and Kikuchi [38], where natural frequency maximization problems are implemented by means of the homogenization-based method. Currently, topology optimization is extensively used to conduct eigenvalue optimization to make sure that the fundamental eigenfrequencies are higher than the disturbance frequencies of the system [82]. Also, natural frequency problems are employed to demonstrate the effectiveness of one algorithm [106]. This section focuses on maximizing the first natural frequency of the structure based on the work by Fu and Rolfe [50]. Thus, the optimization problem is mathematically described by:

$$\begin{aligned} \min : & C(X_e) = -\varphi_1 = -\omega_1^2 \\ \text{subject to} : & (\mathbf{K} - \varphi_1 \mathcal{M})\mathbf{u}_1 = 0, \\ & \frac{\sum_{e=1}^{M} X_e V_e}{\sum_{e=1}^{M} V_e} - V^* \leq 0, \\ & 0 < \rho_{\min} \leq X_e \leq 1;\ e = 1, 2, \cdots, M, \end{aligned} \tag{8.1}$$

where $C(X_e)$ is the objective function, X_e is the elemental volume fraction, φ_1 is the first eigenvalue, ω_1 is the first natural frequency, $\mathbf{u}_1$ is the eigenvector related to φ_1, $\mathbf{K}$ is the global stiffness matrix, $\mathcal{M}$ is the global mass matrix, V_e is the volume of the eth element, V^* is the target volume, and $\rho_{\min}$ is a small value, which is 0.001 here. According to the

DOI: 10.1201/9781032634449-8

Rayleigh quotient, the first eigenvalue is expressed by:

$$\varphi_1 = \frac{\mathbf{u}_1^{\mathrm{T}}\mathbf{K}\mathbf{u}_1}{\mathbf{u}_1^{\mathrm{T}}\mathcal{M}\mathbf{u}_1}. \tag{8.2}$$

The global stiffness matrix $\mathbf{K}$ is:

$$\mathbf{K} = (1 - X_e)\rho_{\min}\mathbf{K}^1 + X_e\mathbf{K}^1, \tag{8.3}$$

where $\mathbf{K}^1$ is the stiffness matrix of solid elements.

The global mass matrix $\mathcal{M}$ is:

$$\mathcal{M} = (1 - X_e)\rho_{\min}\mathcal{M}^1 + X_e\mathcal{M}^1, \tag{8.4}$$

where $\mathcal{M}$ is the mass matrix of solid elements.

The sensitivity analysis is then calculated by:

$$\frac{\partial\varphi_1}{\partial X_e} = \mathbf{u}_1^{\mathrm{T}}\left(\frac{\partial\mathbf{K}}{\partial X_e} - \varphi_1\frac{\partial\mathcal{M}}{\partial X_e}\right)\mathbf{u}_1, \tag{8.5}$$

where

$$\frac{\partial\mathbf{K}}{\partial X_e} \approx (1 - X_e)\left.\frac{\partial\mathbf{K}}{\partial X_e}\right|_{X_e=\rho_{\min}} + X_e\left.\frac{\partial\mathbf{K}}{\partial X_e}\right|_{X_e=1}, \tag{8.6}$$

$$\frac{\partial\mathcal{M}}{\partial X_e} \approx (1 - X_e)\left.\frac{\partial\mathcal{M}}{\partial X_e}\right|_{X_e=\rho_{\min}} + X_e\left.\frac{\partial\mathcal{M}}{\partial X_e}\right|_{X_e=1}. \tag{8.7}$$

8.1.2 Numerical Examples

This section will investigate the three representative cases (pin-pin, fixed-pin, and clamped beam) of natural frequency optimization. The first case is a pin-pin beam whose design domain and boundary conditions are shown in Figure 8.1, where M_s is the concentrated mass. In this case, a mesh of 240×60, a target volume fraction of 0.3 ($V^* = 0.3$), a filter radius of 3 ($r_{\min} = 3$), a concentrated mass of 1×10^{-4} ($M_s = 1 \times 10^{-4}$), and a mass density of 1×10^{-9} ($\rho_M = 1 \times 10^{-9}$) are used. The convergence process and optimized topology of the pin-pin beam case are shown in Figure 8.2. Figure 8.2(a) shows that the optimization process converges at the first eigenvalue of 180 after 289 iterations. The resulting topological configuration is shown in Figure 8.2(b).

The second case is the fixed-pin beam whose design domain and boundary conditions are shown in Figure 8.3. The parameter settings are the same as the pin-pin beam case. The

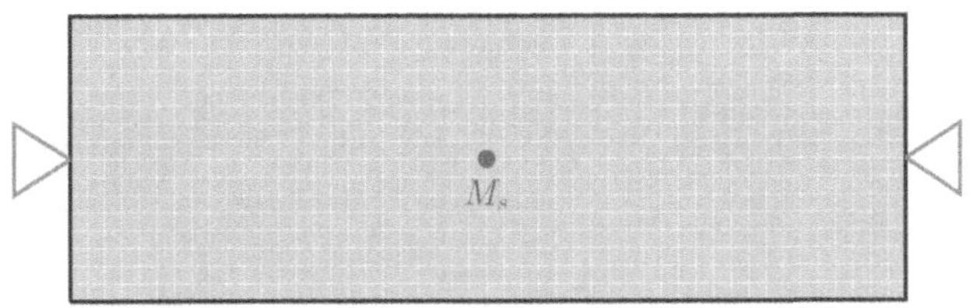

FIGURE 8.1
Design domain of a pin-pin beam.

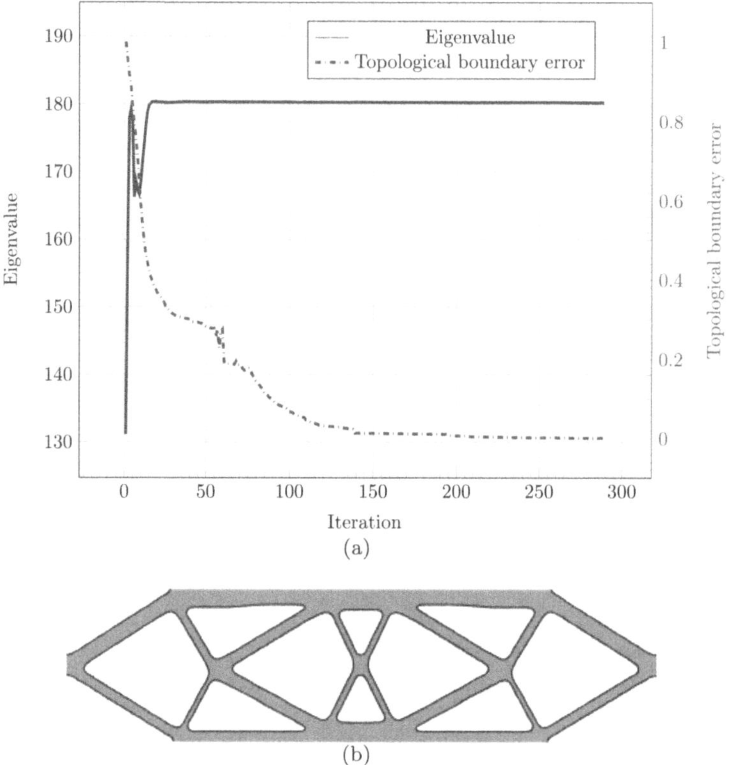

FIGURE 8.2
Pin-pin beam case: (a) convergence process and (b) optimized topology.

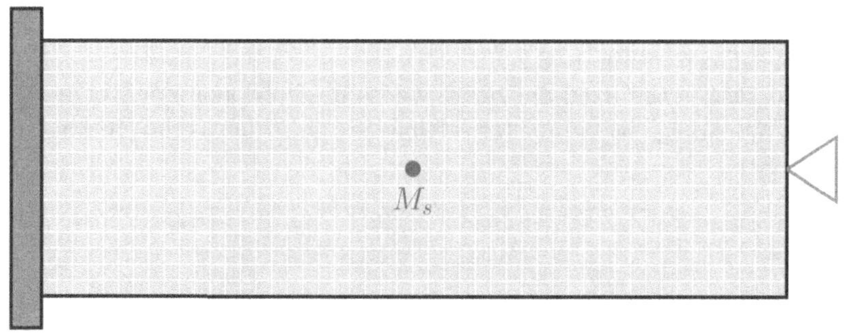

FIGURE 8.3
Design domain of a fixed-pin beam.

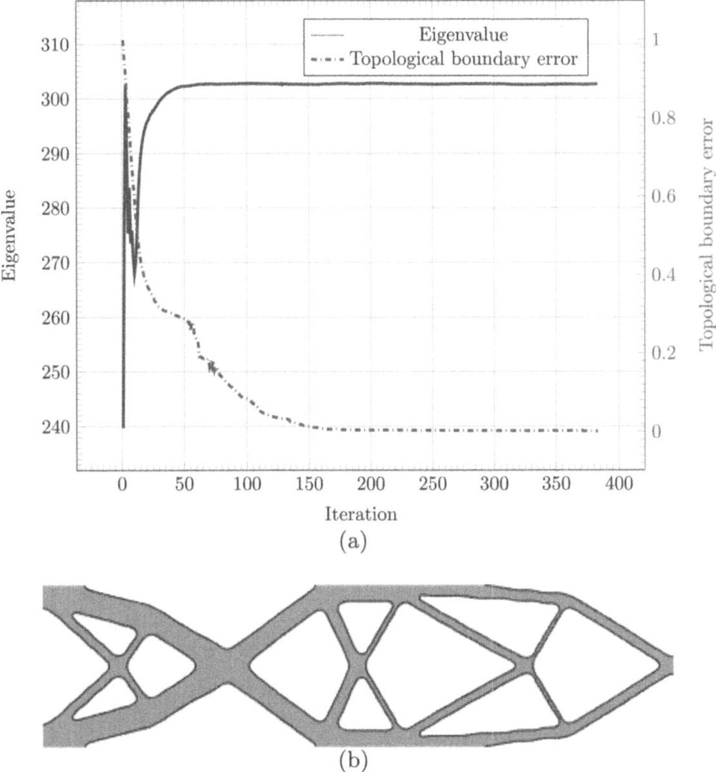

FIGURE 8.4
Fixed-pin beam case: (a) convergence process and (b) optimized topology.

convergence process and optimized topology of the fixed-pin beam case are shown in Figure 8.4. Figure 8.4(a) shows that the optimization process converges at the first eigenvalue of 302 after 383 iterations. The resulting topological configuration is shown in Figure 8.4(b).

The last case is the clamped beam whose design domain and boundary conditions are shown in Figure 8.5. In this case, a mesh of 280×40, $M_s = 1 \times 10^{-3}$, $\rho_M = 1 \times 10^{-9}$, $r_{\min} = 4$, and $V^* = 0.5$ are adopted. The convergence process and optimized topology of the clamped beam case are shown in Figure 8.6. Figure 8.6(a) shows that the optimization process converges at the first eigenvalue of 21.29 after 282 iterations. The resulting topological configuration is shown in Figure 8.6(b).

When a mesh of 1000×200, $M_s = 0.1$, $\rho_M = 1 \times 10^{-9}$, $r_{\min} = 6$, and $V^* = 0.3$ are used, the convergence process and optimized topology are shown in Figure 8.7. Figure 8.7(a) shows that the optimization process converges at the first eigenvalue of 0.2751 after 263 iterations. Figure 8.7(b) shows the resulting topological configuration. The decrease in the target volume fraction from $V^* = 0.5$ to $V^* = 0.3$ leads to a significantly smaller first eigenvalue, and therefore a much less stiff structure. It should also be noted that the concentrate mass has increased by two orders of magnitude in this second clamped case.

Based on the above discussion, it can be concluded that the Smooth-Edged Material Distribution for Optimizing Topology (SEMDOT) algorithm can successfully solve natural frequency maximization problems.

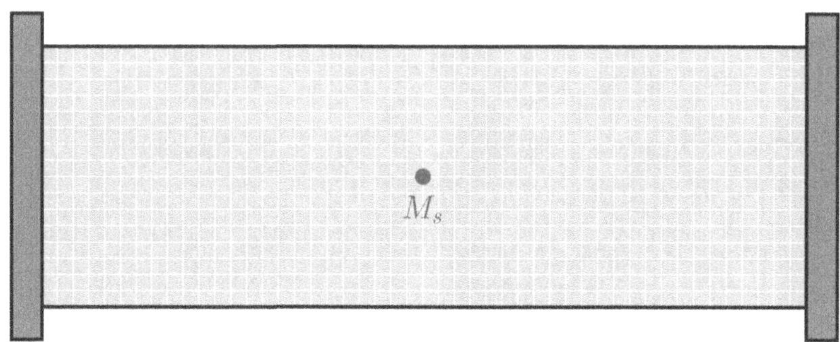

FIGURE 8.5
Design domain of a clamped beam.

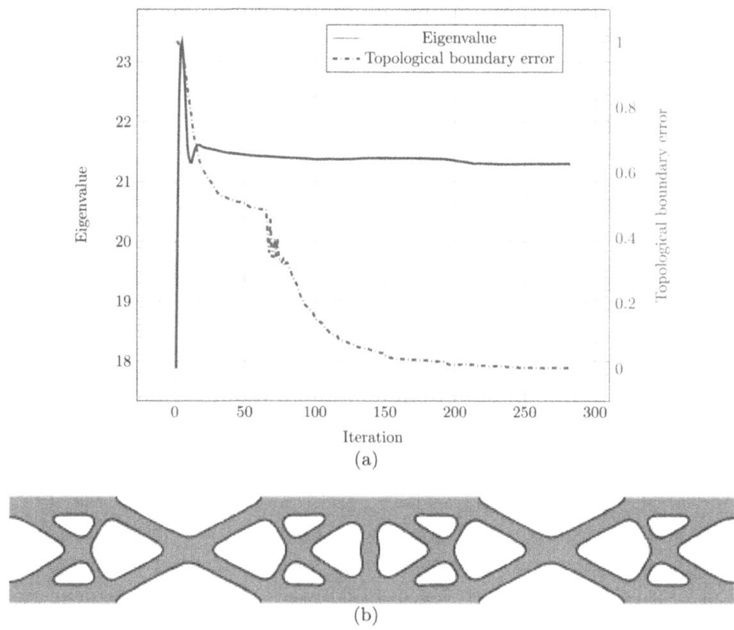

FIGURE 8.6
Clamped beam case with a mesh of 280×40: (a) convergence process and (b) optimized topology.

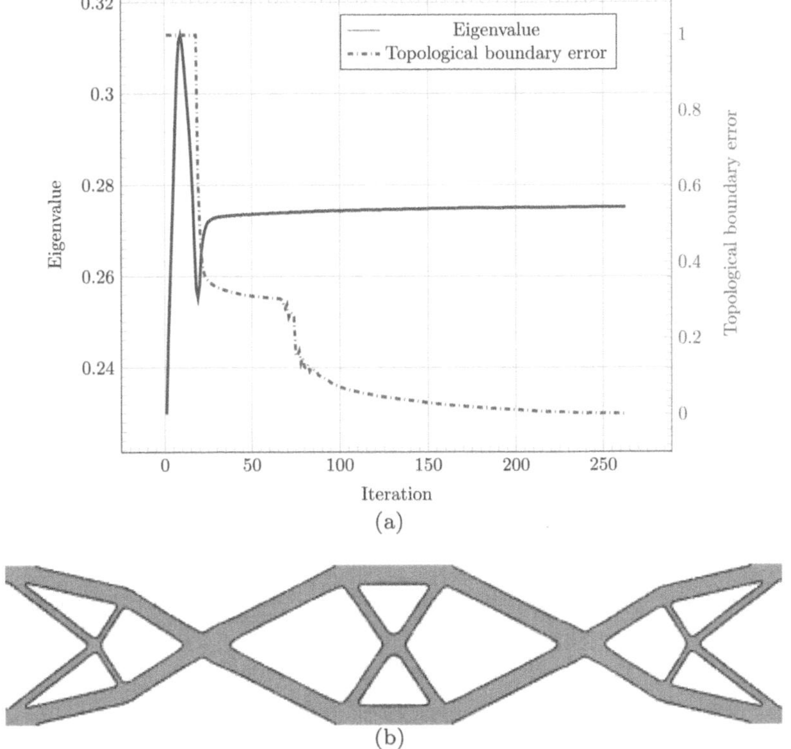

FIGURE 8.7
Clamped beam case with a mesh of 1000×200: (a) convergence process and (b) optimized topology.

8.2 Dynamic Compliance Minimization Problems

8.2.1 Optimization Problem

In many mechanical systems, the sources or drivers of vibration are known, and it is the engineers' task to isolate the system from the vibration. Thus, topology optimization to minimize or maximize the dynamic response of a structure given a driving frequency or range of frequencies is an important industrial problem, also known as a vibration isolation problem. There are numerous approaches to dynamic responses using topology optimization [106, 127, 208, 215]; therefore, this section will concentrate on the work by Niu et al. [135]. Here, dynamic compliance, defined in [137], is taken as an optimization objective to minimize the global response of the vibrating structure:

$$
\begin{aligned}
\min :\ & C_d(X_e) \\
\text{subject to}:\ & \mathbf{K_d u} + \mathbf{D_d \dot{u}} + \mathbf{M_d \ddot{u}} = \mathbf{p}, \\
& \frac{\sum_{e=1}^{M} X_e V_e}{\sum_{e=1}^{M} V_e} - V^* \leq 0, \\
& 0 < \rho_{\min} \leq X_e \leq 1;\ e = 1, 2, \cdots, M,
\end{aligned}
\tag{8.8}
$$

where $C_d(X_e)$ is the objective function for dynamic compliance, $\mathbf{K_d}$ is the global stiffness matrix, $\mathbf{D_d}$ is the damping matrix, $\mathbf{M_d}$ is the mass matrix, $\mathbf{u}$ is the vibration displacement vector, and $\mathbf{p}$ is the time-harmonic external mechanical loading. In addition, $\dot{\mathbf{u}}$ and $\ddot{\mathbf{u}}$ are the first-order and second-order time derivatives of the displacement, namely, the velocity and the acceleration.

The time-harmonic external mechanical loading is calculated by:

$$\mathbf{p}(t) = \mathbf{P_d} e^{i\omega_p t}, \tag{8.9}$$

where $\mathbf{P_d}$ is the prescribed amplitude, ω_p is the frequency, and t is the time.

The vibration displacement is calculated by:

$$\mathbf{u}(t) = \mathbf{U_d} e^{i\omega_p t}, \tag{8.10}$$

where $\mathbf{U_d}$ is the amplitude of the displacement.

The damping can be involved in the standard Rayleigh damping form:

$$\mathbf{D_d} = \alpha_d \mathbf{M_d} + \beta_d \mathbf{K_d}, \tag{8.11}$$

where α_d and β_d are mass-proportional and stiffness-proportional damping coefficients, respectively.

To have the differentiable optimization problem, dynamic compliance is mathematically expressed by:

$$C_d(X_e) = (\mathbf{P_d^T U_d})\overline{(\mathbf{P_d^T U_d})}, \tag{8.12}$$

where the overbar is the conjugate of a complex quantity.

Sensitivities can be calculated by:

$$\frac{\partial C_d(X_e)}{\partial X_e} = \frac{\partial(\mathbf{P_d^T U_d})}{\partial X_e}\overline{(\mathbf{P_d^T U_d})} + (\mathbf{P_d^T U_d})\frac{\partial\overline{(\mathbf{P_d^T U_d})}}{\partial X_e}, \tag{8.13}$$

where

$$\mathbf{U_d} = \mathbf{U_R} + i\mathbf{U_S}, \tag{8.14}$$

$$\mathbf{P_d} = \mathbf{P_R} + i\mathbf{P_S}, \tag{8.15}$$

where i is the imaginary unit; $\mathbf{U_R}$ and $\mathbf{U_S}$ are real and imaginary parts of $\mathbf{U_d}$, respectively; $\mathbf{P_R}$ and $\mathbf{P_S}$ are real and imaginary parts of $\mathbf{P_d}$, respectively.

The governing equation is given by:

$$\mathbf{G_d U_d} = \mathbf{P_d}, \tag{8.16}$$

where

$$\mathbf{G_d} = \mathbf{K_d} + i\omega_p \mathbf{D_d} - \omega_p^2 \mathbf{M_d}, \tag{8.17}$$

where $\mathbf{G_d}$ is the global dynamic stiffness matrix.

Then, the sensitivity analysis can be expressed by:

$$\begin{aligned}\frac{\partial C_d(X_e)}{\partial X_e} &= -2\mathrm{Re}\left[\overline{\mathbf{P_d^T U_d}}\left(\mathbf{U_d^T}\frac{\partial \mathbf{G_d}}{\partial X_e}\mathbf{U_d}\right)\right] \\ &= -2\mathrm{Re}\left[\overline{\mathbf{P_d^T U_d}}\left(\mathbf{U_d^T}\left(\frac{\partial \mathbf{K_d}}{\partial X_e} + i\omega_p\frac{\partial \mathbf{D_d}}{\partial X_e} - \omega_p^2\frac{\partial \mathbf{M_d}}{\partial X_e}\right)\mathbf{U_d}\right)\right],\end{aligned} \tag{8.18}$$

where Re is the real part of the quantity.

Under the condition that $\mathbf{P_d}$ is real, the sensitivity analysis can be reduced as:

$$\frac{\partial C_d(X_e)}{\partial X_e} = -2\mathrm{Re}\left[\mathbf{P_d^T}\overline{\mathbf{U_d}}\left(\mathbf{U_d^T}\frac{\partial \mathbf{G_d}}{\partial X_e}\mathbf{U_d}\right)\right]. \tag{8.19}$$

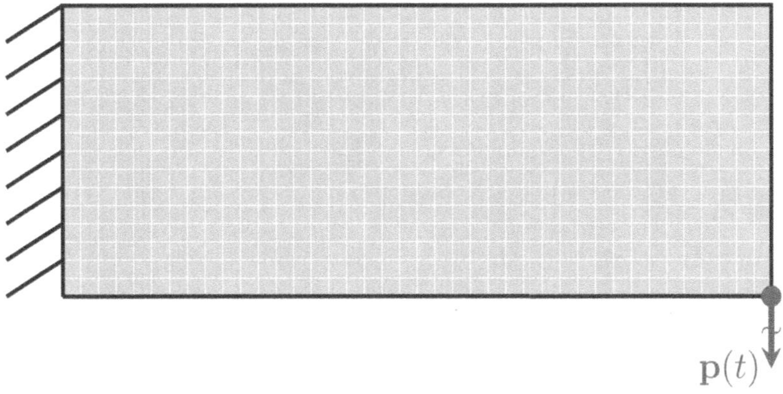

FIGURE 8.8
Design domain of dynamic compliance case.

8.2.2 Numerical Examples

The optimization example is illustrated in Figure 8.8. In this case, the structure consists of two different materials. The stiffer material has the Young's modulus of 7×10^4, Poisson's ratio of 0.3, and mass density of 2.7×10^{-9}. The weaker material has the Young's modulus of 3×10^3, Poisson's ratio of 0.3, and mass density of 1.2×10^{-9}. In addition, the values of the Rayleigh damping coefficients are $\alpha_d = 0.524$ and $\beta_d = 1.63 \times 10^{-7}$. A mesh of 200×60, $r_{\min} = 3$, $V^* = 0.3$, and $\mathbf{P_d} = 1 \times 10^3$ are used here.

The stiffness matrix of the eth element is expressed by:

$$\mathbf{k}_e = X_e\mathbf{k}_e^s + (1 - X_e)\mathbf{k}_e^w, \tag{8.20}$$

where $\mathbf{k}_e^s$ is the stiffness matrix of the stiffer material and $\mathbf{k}_e^w$ is the stiffness matrix of the weaker material.

The mass matrix of the eth element is expressed by:

$$\mathbf{m}_e = X_e\mathbf{m}_e^s + (1 - X_e)\mathbf{m}_e^w, \tag{8.21}$$

where $\mathbf{m}_e^s$ is the mass matrix of the stiffer material and $\mathbf{m}_e^w$ is the mass matrix of the weaker material.

When utilizing a driving frequency of 700 Hz (ω_p = 700 Hz) with a scaling factor of 0.01 for the MMA optimizer, the resulting designs (element-based and smooth) of the stiffer material configurations are shown in Figure 8.9. Figure 8.9(a) shows that the optimization process converges at the dynamic compliance of 6,819 after 151 iterations. The element-based design using two different materials is shown in Figure 8.9(b) where black elements represent the stiffer material, and gray elements represent the weaker material. The smooth design of the stiffer material is shown in Figure 8.9(c).

When considering a driving frequency of 1,500 Hz ($\omega_p = 1,500$ Hz) with a scaling factor of 0.1, the resulting designs (element-based and smooth) are shown in Figure 8.10. Figure 8.10(a) shows that the optimization process converges at the dynamic compliance of 758 after 176 iterations. Figure 8.10(b) shows the element-based design using two different materials. Figure 8.10(c) shows the smooth design of the stiffer material.

In dynamic compliance minimization problems, some parameter settings can result in unreasonable topological configurations. The first example showing this behavior is when the driving frequency is reduced to 50 Hz ($\omega_p = 50$ Hz) with a scaling factor of 1×10^{-6}, and the volume fraction is set to 0.3 ($V^* = 0.3$), the element-based design using two materials

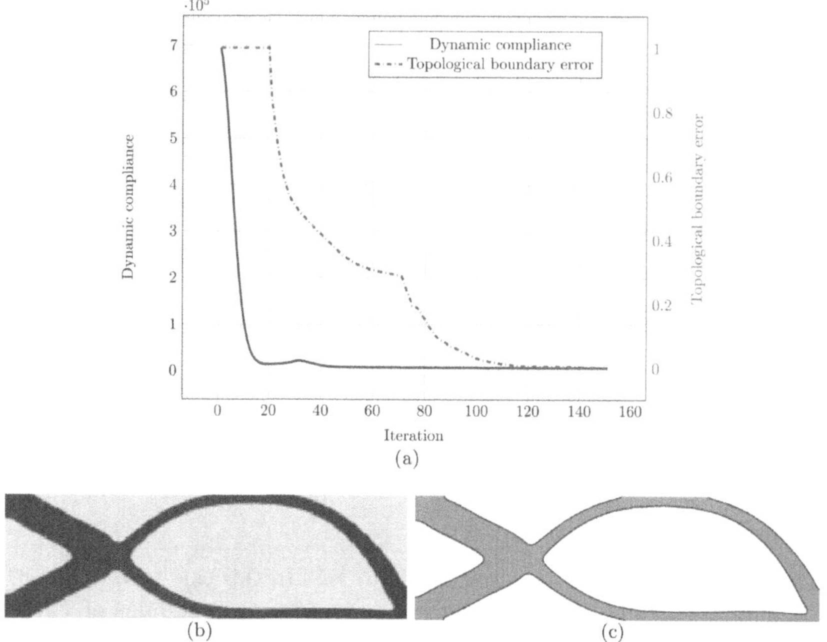

FIGURE 8.9
Dynamic compliance case with a frequency of 700 Hz: (a) optimization process, (b) element-based design, and (c) smooth design.

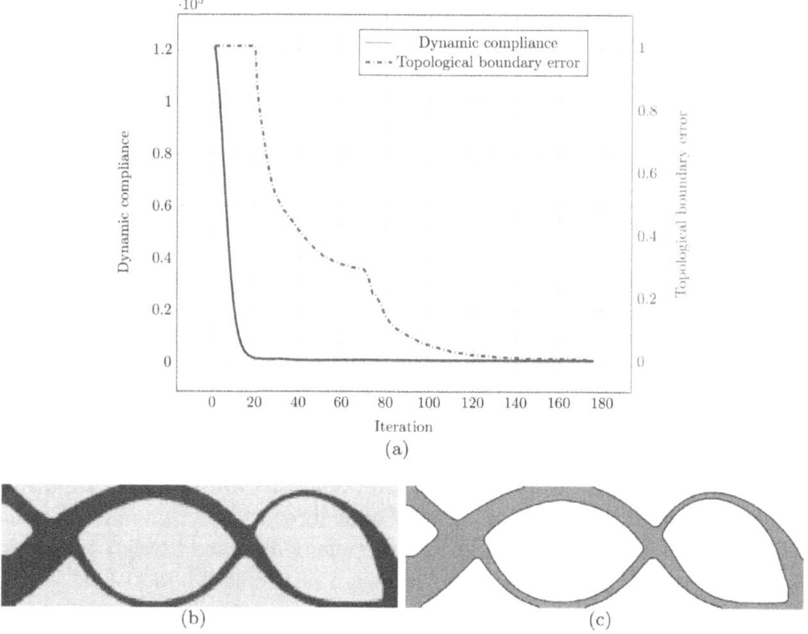

FIGURE 8.10
Dynamic compliance case with a driving frequency of 1500 Hz: (a) optimization process, (b) element-based design, and (c) smooth design.

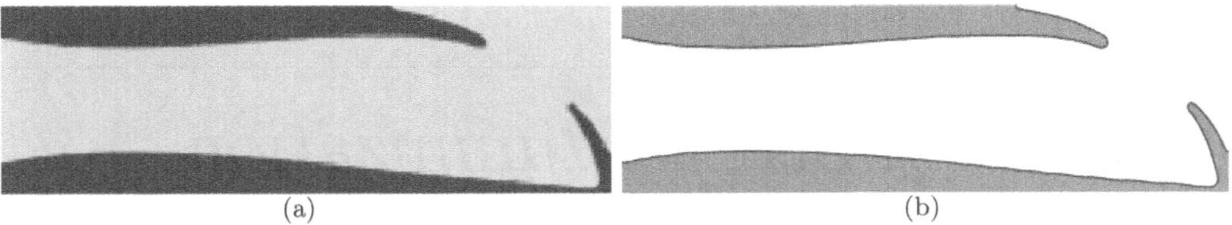

FIGURE 8.11
Element-based design and smooth design subjected to the driving frequency of 50 Hz and $V^* = 0.3$.

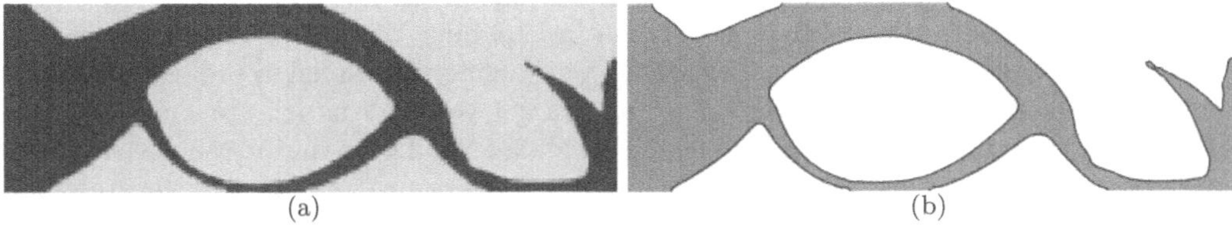

FIGURE 8.12
Element-based design and smooth design subjected to the driving frequency of 1500 Hz and $V^* = 0.4$.

and the smooth design of just the stiffer material are shown in Figure 8.11(a) and 8.11(b), respectively. The structure in Figure 8.11 cannot be regarded as a valid topological design.

The second example showing unreasonable topological configurations is when $\omega_p =$ 1,500 Hz and $V^* = 0.4$ are adopted, the element-based design using two materials and the smooth design of the stiffer material are shown in Figure 8.12(a) and 8.12(b), respectively. Obviously, the structure in Figure 8.12 is not robust nor valid.

Based on the above discussion, it can be concluded that the SEMDOT algorithm can successfully solve dynamic compliance minimization problems; however it should also be admitted that the final results obtained by SEMDOT are very sensitive to the parameter settings.

8.3 Summary

The SEMDOT algorithm successfully solved both natural frequency maximization and dynamic compliance minimization problems, demonstrating its effectiveness in dynamic problems. However, when solving dynamic compliance minimization problems, the authors suggest that the parameters should be selected properly to obtain reasonable or robust topological configurations.

9

Maximum Stress Minimization Problems

This chapter implements topology optimization of maximum stress minimization problems. The L-bracket case is selected to demonstrate the capability of Smooth-Edged Material Distribution for Optimizing Topology (SEMDOT) in solving stress-related problems. Effects of the stress penalty coefficient, filter radius, and target volume fraction are comprehensively discussed. Two strategies of conducting the boundary validation are presented. A new termination criterion is proposed to speed up the convergence process of the stress-based SEMDOT algorithm.

9.1 Optimization Problem

Stress-related problems are extremely important in topology optimization since topological structures that are free of or present less stress concentrations can be obtained [76, 136, 145]. In the literature, p-norm and Kreisselmeier-Steinhauser measures are the most widely used approaches in stress-related topology optimization problems [91, 125, 128, 185]. However, it should be mentioned that existing stress-related approaches are very sensitive to input parameters. In this chapter, the p-norm stress measure is taken as the objective function. Therefore, the maximum stress minimization problem is mathematically stated as:

$$\begin{aligned} \min : C(X_e) = \sigma^{PN} &= \left(\sum_{e=1}^{M}\left(\frac{\sigma_e^{VM}}{\bar{\sigma}}\right)^p\right)^{\frac{1}{p}} \\ \text{subject to} : \mathbf{Ku} &= \mathbf{F}, \\ &\frac{\sum_{e=1}^{M} X_e V_e}{\sum_{e=1}^{M} V_e} - V^* \leq 0, \\ &0 < \rho_{\min} \leq X_e \leq 1;\ e = 1, 2, \cdots, M, \end{aligned} \tag{9.1}$$

where $C(X_e)$ is the objective function, X_e is the elemental volume fraction, p is the aggregation parameter, σ^{PN} is the p-norm stress, σ_e^{VM} is the von Mises stress amplitude of the eth element, $\bar{\sigma}$ is the allowable stress value, $\mathbf{K}$ is the global stiffness matrix, $\mathbf{u}$ is the displacement vector, $\mathbf{F}$ is the load vector, M is the total number of elements, V_e is the volume of the eth element, V^* is the target volume, and $\rho_{\min}$ is a small value, which is 0.001 here. If the aggregation parameter p tends to infinity, σ^{PN} is equivalent to $\max \frac{\sigma_e^{VM}}{\bar{\sigma}}$. However, a high value of p could cause the numerical instability during optimization.

DOI: 10.1201/9781032634449-9

Here, the penalized element stress amplitude vector is given by:

$$\boldsymbol{\sigma_e} = X_e^s \boldsymbol{D}_0 \boldsymbol{B}_c \boldsymbol{u}_e = \{\sigma_{ex}, \sigma_{ey}, \tau_{exy}\}^{\mathrm{T}}, \tag{9.2}$$

where s is the stress penalty coefficient, $\boldsymbol{D}_0$ is the elastic matrix of the solid material, $\boldsymbol{B}_c$ is the strain displacement matrix of the element center, $\boldsymbol{u}_e$ is the displacement vector of the eth element, and σ_{ex}, σ_{ey}, and τ_{exy} are the three stress components of the eth element. To mitigate the singularity phenomenon, the stress penalty coefficient is generally chosen between 0 and 1 (that is, $0 < s < 1$) [23, 72, 98].

The von Mises stress amplitude of the eth element is:

$$\sigma_e^{VM} = \sqrt{\sigma_{ex}^2 + \sigma_{ey}^2 - \sigma_{ex}\sigma_{ey} + 3\tau_{exy}^2}. \tag{9.3}$$

9.2 Sensitivity Analysis

The derivative of the p-norm stress measure with respect to the elemental von Mises stress is expressed by:

$$\frac{\partial \sigma^{PN}}{\partial \sigma_e^{VM}} = \frac{\left[\sum_{e=1}^{M}\left(\frac{\sigma_e^{VM}}{\bar{\sigma}}\right)^p\right]^{\frac{1}{p}-1}\left(\frac{\sigma_e^{VM}}{\bar{\sigma}}\right)^{p-1}}{\bar{\sigma}}. \tag{9.4}$$

The derivatives of the elemental von Mises stress with respect to the three stress components are given by:

$$\frac{\partial \sigma_e^{VM}}{\partial \sigma_{ex}} = \frac{1}{2\sigma_e^{VM}}(2\sigma_{ex} - \sigma_{ey}), \tag{9.5}$$

$$\frac{\partial \sigma_e^{VM}}{\partial \sigma_{ey}} = \frac{1}{2\sigma_e^{VM}}(2\sigma_{ey} - \sigma_{ex}), \tag{9.6}$$

$$\frac{\partial \sigma_e^{VM}}{\partial \tau_{exy}} = \frac{3\tau_{exy}}{\sigma_e^{VM}}. \tag{9.7}$$

The derivative of the penalized element stress amplitude vector with respect to the design variable is calculated by:

$$\frac{\partial \boldsymbol{\sigma_e}}{\partial X_i} = \frac{\partial X_e^s}{\partial X_i}\boldsymbol{D}_0 \boldsymbol{B}_c \boldsymbol{u}_e + X_e^s \boldsymbol{D}_0 \boldsymbol{B}_c \frac{\partial \boldsymbol{u}_e}{\partial X_i}. \tag{9.8}$$

As the term $\frac{\partial \boldsymbol{\sigma_e}}{\partial X_i} \neq 0$ only for $e = i$, Equation (9.8) is therefore rewritten as:

$$\frac{\partial \boldsymbol{\sigma_e}}{\partial X_i} = sX_i^{s-1}\boldsymbol{D}_0 \boldsymbol{B}_c \boldsymbol{u}_i + X_e^s \boldsymbol{D}_0 \boldsymbol{B}_c \frac{\partial \boldsymbol{u}_e}{\partial X_i}. \tag{9.9}$$

According to $\mathbf{Ku} = \mathbf{F}$, we have:

$$\frac{\partial \boldsymbol{u}}{\partial X_i} = \mathbf{K}^{-1}\left(\frac{\partial \mathbf{F}}{\partial X_i} - \frac{\partial \mathbf{K}}{\partial X_i}\mathbf{u}\right). \tag{9.10}$$

Substituting (9.10) into (9.9) yields:

$$\frac{\partial \boldsymbol{\sigma}_e}{\partial X_i} = sX_i^{s-1}\boldsymbol{D}_0\boldsymbol{B}_c\boldsymbol{u}_i + X_i^s\boldsymbol{D}_0\boldsymbol{B}_c\mathbf{K}^{-1}\left(\frac{\partial \mathbf{F}}{\partial X_i} - \frac{\partial \mathbf{K}}{\partial X_i}\mathbf{u}\right). \tag{9.11}$$

With the following adjoint problem, we have:

$$\mathbf{K}\boldsymbol{\lambda} = \sum_{e=1}^{M}\frac{\partial \sigma^{PN}}{\partial \sigma_e^{VM}}\boldsymbol{B}_c^{\mathrm{T}}\boldsymbol{D}_0^{\mathrm{T}}\left(\frac{\partial \sigma_e^{VM}}{\partial \sigma_e}\right). \tag{9.12}$$

Therefore, the derivative of the p-norm stress measure with respect to the design variable can be calculated by:

$$\frac{\partial \sigma^{PN}}{\partial X_i} = \frac{\partial \sigma^{PN}}{\partial \sigma_i^{VM}}\left(\frac{\partial \sigma_i^{VM}}{\partial \boldsymbol{\sigma}_i}\right)^{\mathrm{T}} sX_i^{s-1}\boldsymbol{D}_0\boldsymbol{B}_c\boldsymbol{u}_i - X_i^s\lambda_i^{\mathrm{T}}\frac{\partial \mathbf{K}_i}{\partial X_i}\boldsymbol{u}_i. \tag{9.13}$$

To follow the Smooth-Edged Material Distribution for Optimizing Topology (SEMDOT) form, Equation (9.13) is rewritten as:

$$\frac{\partial \sigma^{PN}}{\partial X_i} = (X_e + (1 - X_e)\rho_{\min})\left[\frac{\partial \sigma^{PN}}{\partial \sigma_i^{VM}}\left(\frac{\partial \sigma_i^{VM}}{\partial \boldsymbol{\sigma}_i}\right)^{\mathrm{T}} sX_i^{s-1}\boldsymbol{D}_0\boldsymbol{B}_c\boldsymbol{u}_i - X_i^s\lambda_i^{\mathrm{T}}\mathbf{K}_i^1\boldsymbol{u}_i\right], \tag{9.14}$$

where $\mathbf{K}_i^1$ is the stiffness matrix of the solid element.

9.3 Numerical Examples

9.3.1 Effects of Varying Stress Penalty Coefficient

The L-bracket beam shown in Figure 9.1 is used to demonstrate the capability of SEMDOT in solving the stress minimization problems. As depicted in Figure 9.1, to avoid the stress concentration, the load is evenly distributed on six neighboring nodes. A mesh of 100×100, filter radius of 4 ($r_{\min}$=4), target volume fraction of 0.3 ($V^* = 0.3$), aggregation parameter

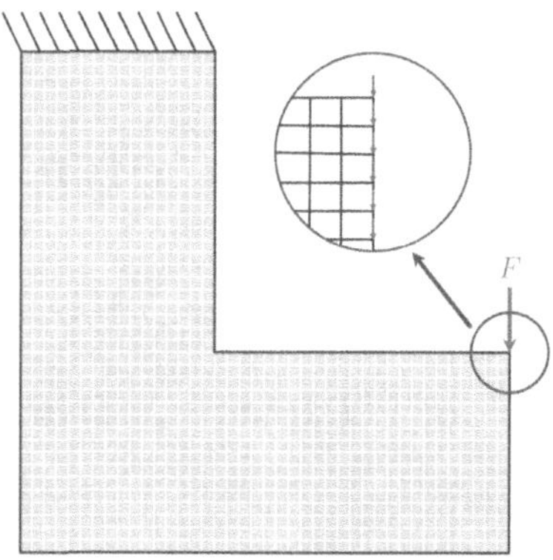

FIGURE 9.1
Design domain of L-bracket beam.

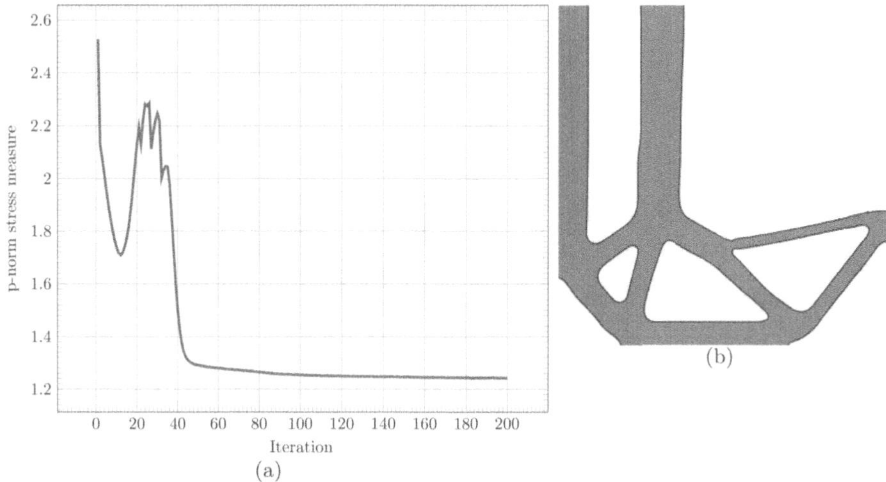

FIGURE 9.2
L-bracket case subjected to $s = 0.4$: (a) optimization process and (b) optimized topology.

of 10 ($p = 10$), and allowable stress value of 0.55 ($\bar{\sigma} = 0.55$) are used. In addition, the move limit in Method of Moving Asymptote (MMA) is set to 1. To stabilize the optimization process, a positive move limit is introduced in the MMA solver. For each iterative step, the elemental volume fraction X_e meets:

$$\max(X_e - MOV, \rho_{\min}) \leq X_e \leq \min(X_e + MOV, 1), \tag{9.15}$$

where MOV is the positive move limit, which is set to 0.01 here.

When solving stress-based problems, it is difficult to meet the convergence criteria in SEMDOT. The maximum number of iterations of 200 is adopted for the purpose of terminating the optimization process. Therefore, only the performance will be discussed here. When using the stress penalty coefficient of 0.4 (that is, $s = 0.4$), the optimization process and optimized topology are shown in Figure 9.2. In Figure 9.2(a), the final p-norm stress value is 1.242. Figure 9.2(b) shows that a curvature is formed at the reentrant corner to reduce the stress concentration.

The p-norm stress values under different stress penalty coefficients (that is, s=0.01, 0.1, 0.25, 0.5, 0.55, and 0.6) are summarized in Table 9.1. With the rise in the stress penalty coefficient, the p-norm stress value increases. Optimized topologies are shown in Figure 9.3. Obviously, a high value of the stress penalty coefficient (for example, s=0.55 and 0.6) can result in unacceptable topological configurations. In addition, a high value of the stress penalty coefficient can lead to a big curvature at the reentrant corner, which is beneficial in terms of reducing the stress concentration.

TABLE 9.1
p-norm stress measure subjected to different stress penalty coefficients

Stress penalty coefficient s	**0.01**	**0.1**	**0.25**	**0.5**	**0.55**	**0.6**
p-norm stress measure	7.761	4.189	1.508	1.804	1.484	1.251

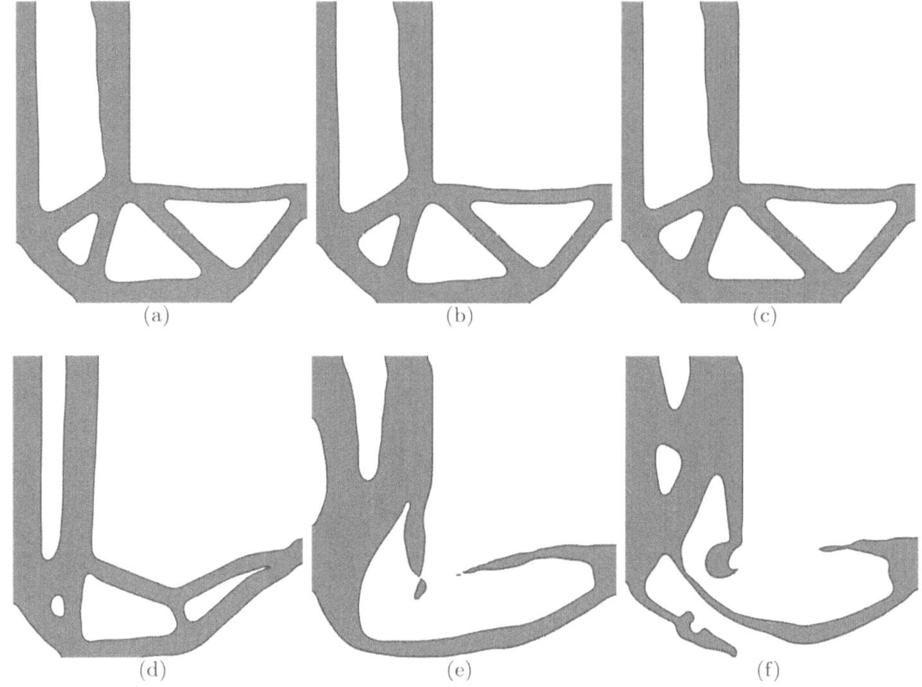

FIGURE 9.3
Resulting topologies of L-bracket case under different stress penalty coefficients: (a) $s = 0.01$, (b) $s = 0.1$, (c) $s = 0.25$, (d) $s = 0.5$, (e) $s = 0.55$, and (f) $s = 0.6$.

9.3.2 Effects of Varying Filter Radius

The p-norm stress values under different filter radii (that is, $r_{\min}$=1.5, 2.5, 3.5, 4.5, 5.5, and 6.5) are summarized in Table 9.2. It is noted that $V^* = 0.3$ is adopted here. The effect of the filter radius on the p-norm stress measure is not significant. Optimized topologies are shown in Figure 9.4. Different filter radii can lead to distinct topological configurations. A small value of the filter radius (for example, $r_{\min}$=1.5 and 2.5) can lead to discontinuous structures.

9.3.3 Effects of Varying Volume Fractions

The p-norm stress values under different target volume fractions (that is, V^*=0.15, 0.2, 0.4, and 0.5) are outlined in Table 9.3. It is noted that $r_{\min} = 4$ is employed here. Except for $V^* = 0.4$, with the increase in the target volume fraction, the p-norm stress measure

TABLE 9.2
p-norm stress measure subjected to different filter radii

Filter radius $r_{\min}$	**1.5**	**2.5**	**3.5**	**4.5**	**5.5**	**6.5**
p-norm stress measure	1.265	1.329	1.247	1.245	1.245	1.244

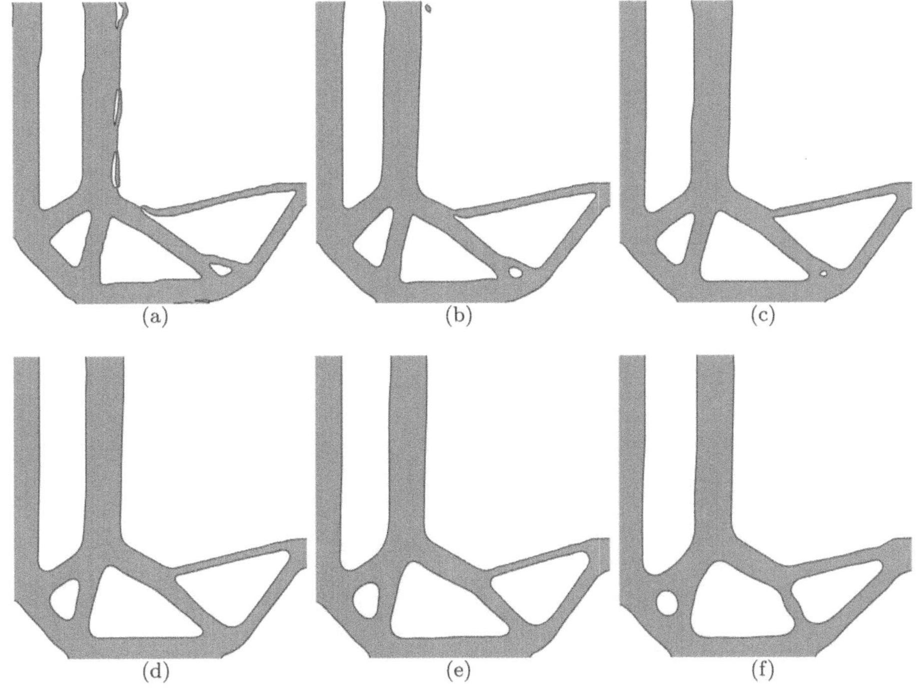

FIGURE 9.4
Resulting topologies of L-bracket case under different filter radii: (a) $r_{\min} = 1.5$, (b) $r_{\min} = 2.5$, (c) $r_{\min} = 3.5$, (d) $r_{\min} = 4.5$, (e) $r_{\min} = 5.5$, and (f) $r_{\min} = 6.5$.

decreases. The corresponding optimized topologies are shown in Figure 9.5. If value of the volume fraction is small (for example, V^*=0.15), a proper structure still can be formed without any discontinuous boundary issue.

9.4 Remarks on Allowable Stress Value

The p-norm stress values under different allowable stresses (that is, $\bar{\sigma}$=0.35, 0.65, 0.8, and 1.2) are outlined in Table 9.4. It is noted that $V^* = 0.3$ and $r_{\min} = 4$ are considered here. It is apparent that the allowable stress can significantly affect the p-norm stress measure. However, the p-norm stress value does not mean the real stress distribution. Resulting topologies are shown in Figure 9.6, where the same topological configurations are observed.

TABLE 9.3
p-norm stress measure subjected to different target volume fractions

Target volume fraction V^*	**0.15**	**0.2**	**0.4**	**0.5**
p-norm stress measure	1.921	1.534	1.193	1.242

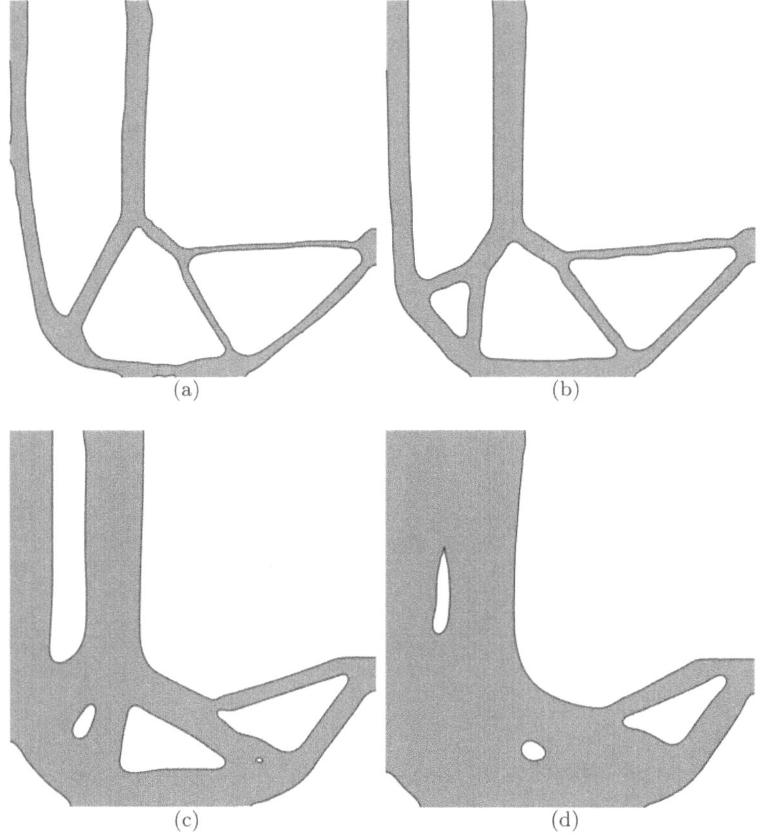

FIGURE 9.5
Resulting topologies of L-bracket case under different target volume fractions: (a) $V^* = 0.15$, (b) $V^* = 0.2$, (c) $V^* = 0.4$, and (d) $V^* = 0.5$.

This is because the stress is used as the objective not constraint. However, stress-constrained problems can also be solved by extending the work in this chapter. Some existing methods such as the augmented Lagrangian method [35, 36, 62, 211] can be used for this purpose. If the stress is taken as the constraint, the topological configuration will vary with the allowable stress. The stress constraint can be expressed by:

$$\sigma^{PN} \leq 1. \tag{9.16}$$

TABLE 9.4
p-norm stress measure subjected to different allowable stress values

Allowable stress value $\bar{\sigma}$	**0.35**	**0.65**	**0.8**	**1.2**
p-norm stress measure	1.960	1.052	0.856	0.567

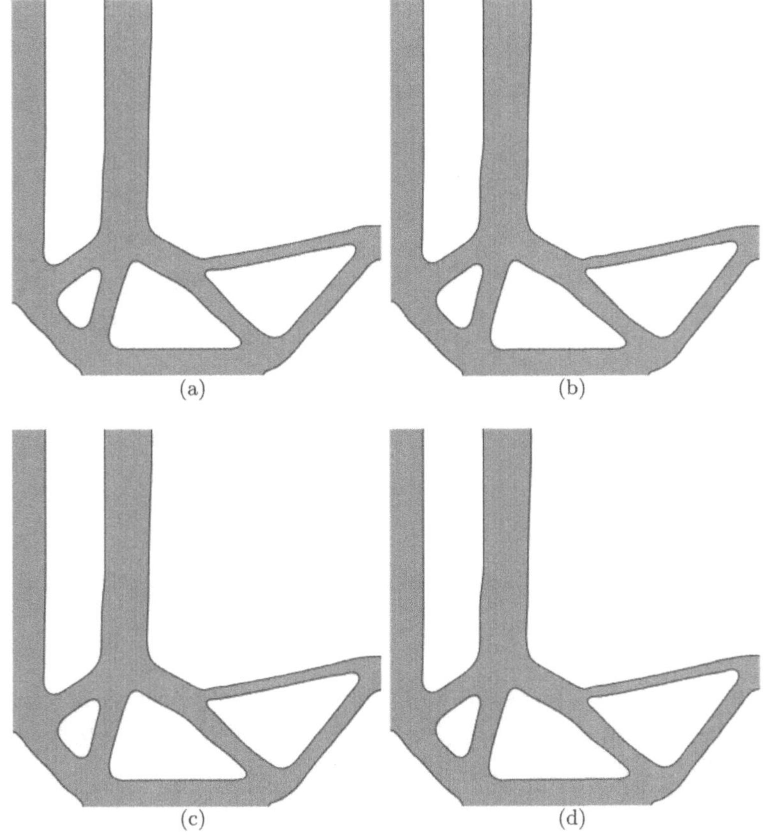

FIGURE 9.6
Resulting topologies of L-bracket case under different allowable stresses: (a) $\bar{\sigma} = 0.35$, (b) $\bar{\sigma} = 0.65$, (c) $\bar{\sigma} = 0.8$, and (d) $\bar{\sigma} = 1.2$.

9.5 A New Termination Criterion

To address the convergence issue in the stress-based SEMDOT algorithm, a new termination criterion named the solidification value is given by:

$$1 - \frac{\sum_{e=1}^{M} X_e^k (1 - X_e^k)}{M} \geq \tau^{\text{stress}}, \tag{9.17}$$

where τ^{stress} is the tolerance value for stress-based SEMDOT, which is 0.95. An example with a mesh of 100×100, $r_{\min} = 4$, and $V^* = 0.3$ is shown in Figure 9.7. It is noted that, in the example, only the termination criterion in Equation (9.17) is adopted. A higher solidification value means a better topological design. Figure 9.7(a) shows that the optimization process converges at the p-norm stress measure of 1.256 after 93 iterations. By contrast, when using old termination criteria, the optimization process cannot converge after 1,000 iterations. In addition, a good topological design is obtained, as shown in Figure 9.7(b). Therefore, it can be concluded that the new termination criterion is suitable for the stress-

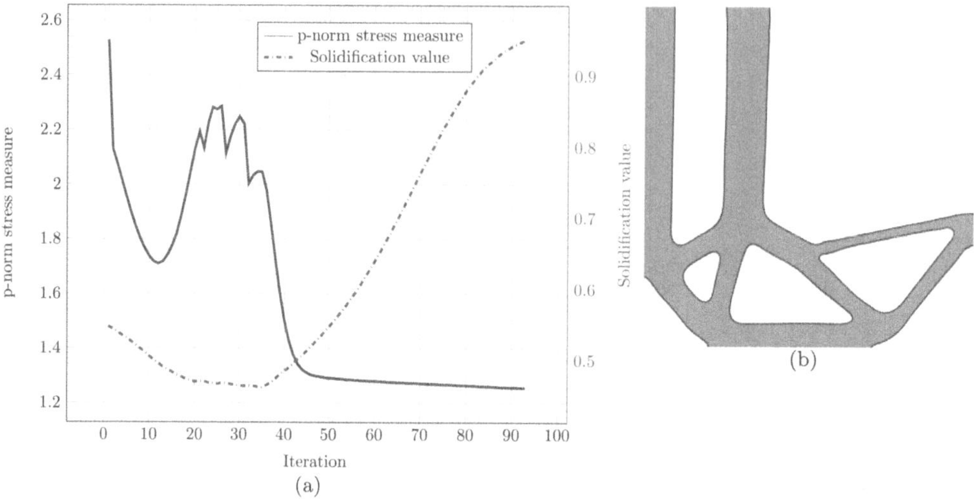

FIGURE 9.7
L-bracket case using a new termination criterion: (a) optimization process and (b) optimized topology.

based SEMDOT algorithm. It should be mentioned that the new termination criterion in SEMDOT is inspired by the one used in the commercial topology optimization software package LS-TaSC.

Although the termination criterion in Equation (9.17) is proposed for stress-based SEMDOT, it can also be used in the optimization problems that are not easy to get converged. Generally, the new termination criterion (that is, the solidification value) requires a shorter convergence process compared to the two old termination criteria (that is, the overall topological alteration and topological boundary error). The selection of the termination strategy should be based on the specific optimization problem.

9.6 Boundary Validation

As the stress value is very sensitive to the boundary, the accurate calculation of the stress distribution based on smooth design is crucial in SEMDOT. There are two strategies that can be used for the boundary validation. The first strategy is that square elements are used for the main body, and arbitrary elements are only used for boundaries, as illustrated in Figure 9.8(a). The second strategy is that the triangulation mesh is adopted for the whole smooth design, as demonstrated in Figure 9.9(a). As shown in Figures 9.8(b) and 9.9(b), von Mises stress distributions of these two strategies are very similar.

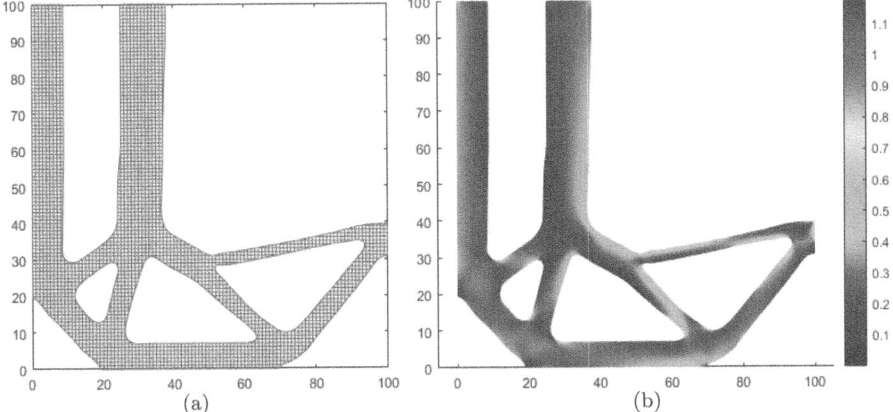

(a) (b)

FIGURE 9.8
Strategy 1 for boundary validation: (a) mesh and (b) von Mises stress.

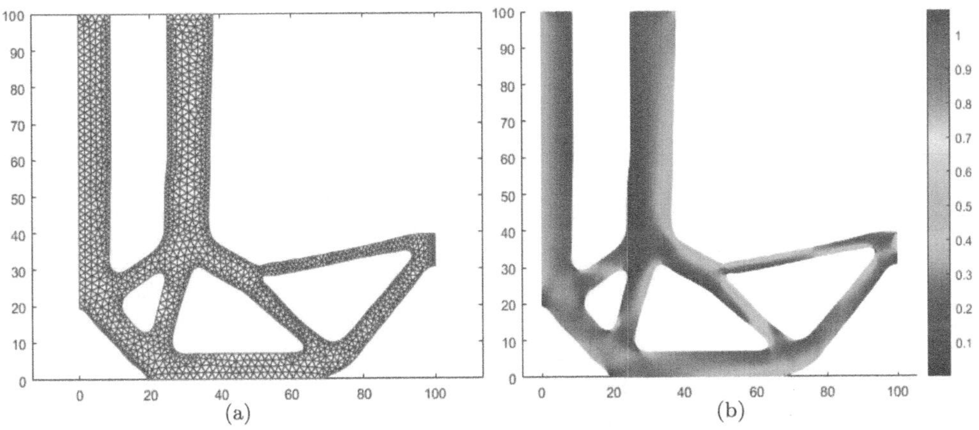

(a) (b)

FIGURE 9.9
Strategy 2 for boundary validation: (a) mesh and (b) von Mises stress.

9.7 Summary

The SEMDOT algorithm successfully solved stress minimization problems. The stress penalty coefficient of 0.4 is recommended for SEMDOT. A proper filter radius can guarantee a reasonable topological configuration. The SEMDOT approach still can obtain an intact topological structure when the target volume fraction is small. Similarly, von Mises stress distributions are observed in two different boundary validation strategies. The work presented in this chapter can be easily extended to stress-constrained problems. To accelerate the convergence process, a new termination criterion named the solidification value is proposed for the stress-based SEMDOT algorithm.

10

Self-Supporting Design for Additive Manufacturing

This chapter combines the Smooth-Edged Material Distribution for Optimizing Topology (SEMDOT) algorithm and Langelaar's Additive Manufacturing (AM) filter to implement self-supporting design, which exemplifies the applications of current topology optimization techniques for AM. Effects of the volume fraction, mesh size, build orientation, and parameter for the approximation smoothness in Langelaar's AM filter on self-supporting design are thoroughly investigated. In addition, a large-scale self-supporting design case is tested to further demonstrate the effectiveness of this combination of algorithms. At the end of this chapter, an optimization of industrial frame case is presented to show the capability of this combination in solving real-world problems.

10.1 Introduction to Design for Additive Manufacturing

The structures obtained by topology optimization are geometrically complex, and hence traditional manufacturing methods such as casting and machining cannot be easily used to fabricate topologically optimized structures [45]. Nowadays, Additive Manufacturing (AM) is extensively used to fabricate the topologically optimized structures [15,84]. Although AM is a very powerful manufacturing technique, it still suffers from some limits [60,86]. Therefore, manufacturing constraints in AM should be incorporated into topology optimization to increase the manufacturability of optimized structures. Topology optimization for AM mainly involves support structure minimization design [31,142,148], self-supporting structure design [21,57,94,146], optimization of the build orientation [95,138,179], feature size control [116], and unsupported or self-supporting enclosed void design [120,174,198,217].

The self-supporting design is taken as an example to demonstrate how to conduct topology optimization for AM by means of Smooth-Edged Material Distribution for Optimizing Topology (SEMDOT). Different self-supporting design methodologies are proposed to solve the overhang angle issue in AM. Figure 10.1 illustrates the limit on the overhang angle in AM. If the angle of the downward facing (overhang) surface of a part is less than the critical overhang angle (a typical value of 45°), the part would deform or warp, resulting in a failure in fabrication (see Figure 10.2). The traditional way of overcoming the overhang angle issue is to introduce the support structures during fabrication [87]. However, this method can significantly increase the use of materials and cost. After printing, extra effort has to be made to remove the support structures. By contrast, self-supporting design approaches do not need neither additional support materials nor removal of support structures. It should be mentioned that the structural performance of self-supporting design structures are sacrificed so the part can be produced without support structures. Nevertheless, in some multi-axis AM technologies, support structures or self-supporting design may not be necessary [24].

DOI: 10.1201/9781032634449-10

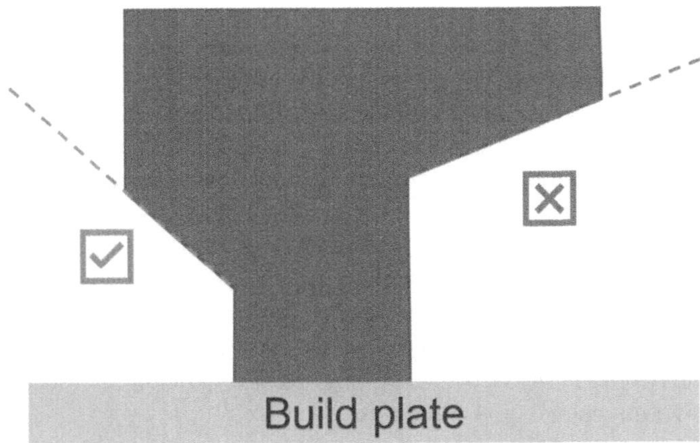

FIGURE 10.1
Limit on overhang angle in additive manufacturing.

FIGURE 10.2
Fabrication failure caused by limit on overhang angle.

10.2 Optimization Problem

Since the Langelaar's AM filter is the most widely used self-supporting design approach [14, 147, 194], and its effectiveness has been fully validated [216], it is selected to combine with SEMDOT to generate smooth support-free parts here. The contents of this chapter are mainly based on the authors' previous work [51,52,54]. As the 3D self-supporting design is more practical from the engineering point of view, this chapter only focuses on the 3D cases.

The fundamental idea of Langelaar's AM filter is that the value of the element to be supported should be less than the maximum value of the elements in the supporting region [93–96]. Langelaar's AM filter is schematically illustrated in Figure 10.3, where the top element is the element to be supported and the five bottom elements ($S_{(i,j,k)}$) form the supporting region. Here, a critical overhang angle of 45° is taken into account. Mathematical expressions of 3D Langelaar's AM filter are:

$$\boldsymbol{\xi}_{(i,j,k)} = \min\left(\boldsymbol{X}_{(i,j,k)}, \boldsymbol{\Xi}_{(i,j,k)}\right) \quad \text{with} \tag{10.1}$$

$$\boldsymbol{\Xi}_{(i,j,k)} = \max\left(\boldsymbol{\xi}_{(i-1,j,k-1)}, \boldsymbol{\xi}_{(i,j-1,k-1)}, \boldsymbol{\xi}_{(i,j,k-1)}, \boldsymbol{\xi}_{(i,j+1,k-1)}, \boldsymbol{\xi}_{(i+1,j,k-1)}\right), \tag{10.2}$$

where $\boldsymbol{X}$ is the vector of elemental volume fractions, $\boldsymbol{\xi}$ is the vector of printed elemental volume fractions, and $\boldsymbol{\Xi}$ is the vector of the maximum printed elemental volume fractions.

To implement gradient-based optimization, the approximation is made as:

$$\boldsymbol{\xi} = \frac{1}{2}\left(\tilde{\mathbf{X}} + \boldsymbol{\Xi} - \left(\left(\tilde{\mathbf{X}} - \boldsymbol{\Xi}\right)^2 + \varepsilon\right)^{\frac{1}{2}} + \sqrt{\varepsilon}\right), \tag{10.3}$$

where $\tilde{\mathbf{X}}$ is the vector of filtered elemental volume fractions and ε is the parameter that controls the accuracy of the approximation.

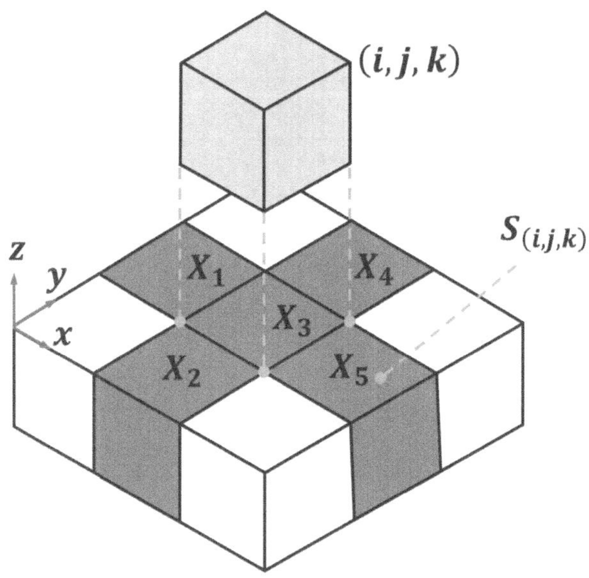

FIGURE 10.3
Illustration of Langelaar's additive manufacturing filter.

The P-Q max function is used to calculate the maximum value of the elements in the supporting region. The P-Q max function is expressed by:

$$\mathrm{smax}(\boldsymbol{\xi}_k) = \boldsymbol{\Xi}(\boldsymbol{\xi}_k) = \left(\sum_{k=1}^{n_s} \boldsymbol{\xi}_k^P\right)^{\frac{1}{Q}}, \tag{10.4}$$

where

$$Q = P + \frac{\log n_s}{\log \xi_0}, \tag{10.5}$$

where $\boldsymbol{\xi}_k$ is the printed elemental volume fraction in the supporting region relevant to the considered element, P is the parameter that controls the approximation smoothness, and n_s is the number of elements in the supporting region. It is noted that $Q \leq P$. A lower value of Q generally results in a stronger penalization on lower elemental volume fractions. The support capability of uniform printed layers below ξ_0 in Equation (10.5) is underestimated by Equation (10.4) [94]. In contrast, the support capability is slightly overestimated when printed elemental volume fractions are above ξ_0 [94].

10.3 Sensitivity Analysis

Here, the stiffness maximization or compliance minimization problem is considered as the optimization objective. The sensitivity of the objective function $C_p(\boldsymbol{\xi})$ with respect to elemental volume fractions $\mathbf{X}$ is calculated with the chain rule as:

$$\frac{\partial C_p(\boldsymbol{\xi})}{\partial \mathbf{X}} = \frac{\partial C_p(\boldsymbol{\xi})}{\partial \boldsymbol{\xi}} \frac{\partial \boldsymbol{\xi}}{\partial \tilde{\mathbf{X}}^{\mathrm{new}}} \frac{\partial \tilde{\mathbf{X}}^{\mathrm{new}}}{\partial \tilde{\mathbf{X}}} \frac{\partial \tilde{\mathbf{X}}}{\partial \mathbf{X}}, \tag{10.6}$$

where $\tilde{\mathbf{X}}^{\mathrm{new}}$ is the vector of new filtered elemental volume fractions.

Using the multipliers, sensitivities of C_p are calculated by:

$$\frac{\partial C_p(\boldsymbol{\xi})}{\partial \tilde{\mathbf{X}}_j} = \boldsymbol{\lambda}_j^{\mathrm{T}} \frac{\partial \boldsymbol{\xi}_j}{\partial \tilde{\mathbf{X}}_j}, \tag{10.7}$$

where $\boldsymbol{\lambda}_j$ is the adjoint field defined as:

$$\begin{aligned} \boldsymbol{\lambda}_j^{\mathrm{T}} &= \frac{\partial C_p(\boldsymbol{\xi})}{\partial \boldsymbol{\xi}_j} + \boldsymbol{\lambda}_{j+1}^{\mathrm{T}} \frac{\partial \boldsymbol{\xi}_{j+1}}{\partial \boldsymbol{\xi}_j}, \quad \text{for } 1 \leq j < n_i, \\ \boldsymbol{\lambda}_{n_i}^{\mathrm{T}} &= \frac{\partial C_p(\boldsymbol{\xi})}{\partial \boldsymbol{\xi}_{n_i}}, \end{aligned} \tag{10.8}$$

The remaining derivatives for sensitivity analysis are calculated by:

$$\frac{\partial \boldsymbol{\xi}}{\partial \tilde{\mathbf{X}}} = \frac{1}{2}\left(1 - \left(\tilde{\mathbf{X}} - \boldsymbol{\Xi}\right)\left(\left(\tilde{\mathbf{X}} - \boldsymbol{\Xi}\right)^2 + \varepsilon\right)^{-\frac{1}{2}}\right), \tag{10.9}$$

$$\frac{\partial \boldsymbol{\xi}}{\partial \boldsymbol{\Xi}} = \frac{1}{2}\left(1 + \left(\tilde{\mathbf{X}} - \boldsymbol{\Xi}\right)\left(\left(\tilde{\mathbf{X}} - \boldsymbol{\Xi}\right)^2 + \varepsilon\right)^{-\frac{1}{2}}\right), \tag{10.10}$$

$$\frac{\partial \Xi_{ki}}{\partial \boldsymbol{\xi}_k} = \frac{P\boldsymbol{\xi}_k^{P-1}}{Q}\left(\sum_{ki=1}^{n_s} \boldsymbol{\xi}_{ki}^P\right)^{\frac{1}{Q}-1}. \tag{10.11}$$

More details on sensitivity analysis can be found in [94].

10.4 Numerical Examples

10.4.1 Effects of Varying Volume Fraction

Firstly, effects of the volume fraction are studied with a simply supported beam shown in Figure 10.4, where the build orientation is in the $+y$ direction. A mesh of $100 \times 50 \times 20$, target volume fraction of 0.3 ($V^* = 0.3$), and filter radius of 4 ($r_{\min} = 4$) are taken into account. To stabilize the convergence process, the move limit in Method of Moving Asymptote (MMA) is set to 0.1, namely, $mov = 0.1$. Optimization processes of general and self-supporting topology optimization and their topologies are shown in Figure 10.5. Figure 10.5(a) shows that the support-free design requires a longer convergence process, and the converged objective values are very close in this case. Obviously, the top corner of the general design in Figure 10.5(b) disobeys the critical overhang angle of 45°. In contrast, the top corner of the self-supporting design in Figure 10.5(c) obeys the critical overhang angle of 45°.

Structural performance and convergence of self-supporting structures for a simply supported beam under a range of target volume fractions (that is, V^*=0.2, 0.4, 0.5, 0.6, 0.7, and 0.8) are summarized in Table 10.1. With the increase in the volume fraction, structural compliance declines. A small target volume fraction can result in a long convergence process. The corresponding topologies are shown in Figure 10.6, where all the designs satisfy the critical overhang angle of 45°.

To further demonstrate the challenge caused by the small target volume fraction, gen-

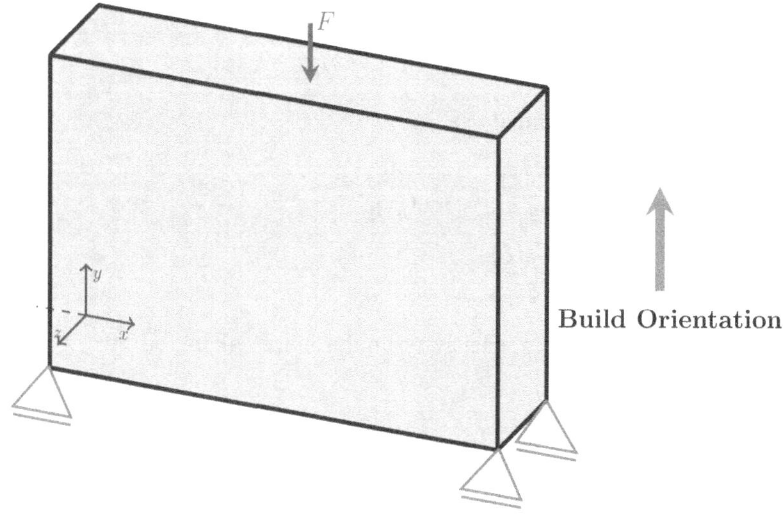

FIGURE 10.4
Design domain of a simply supported beam.

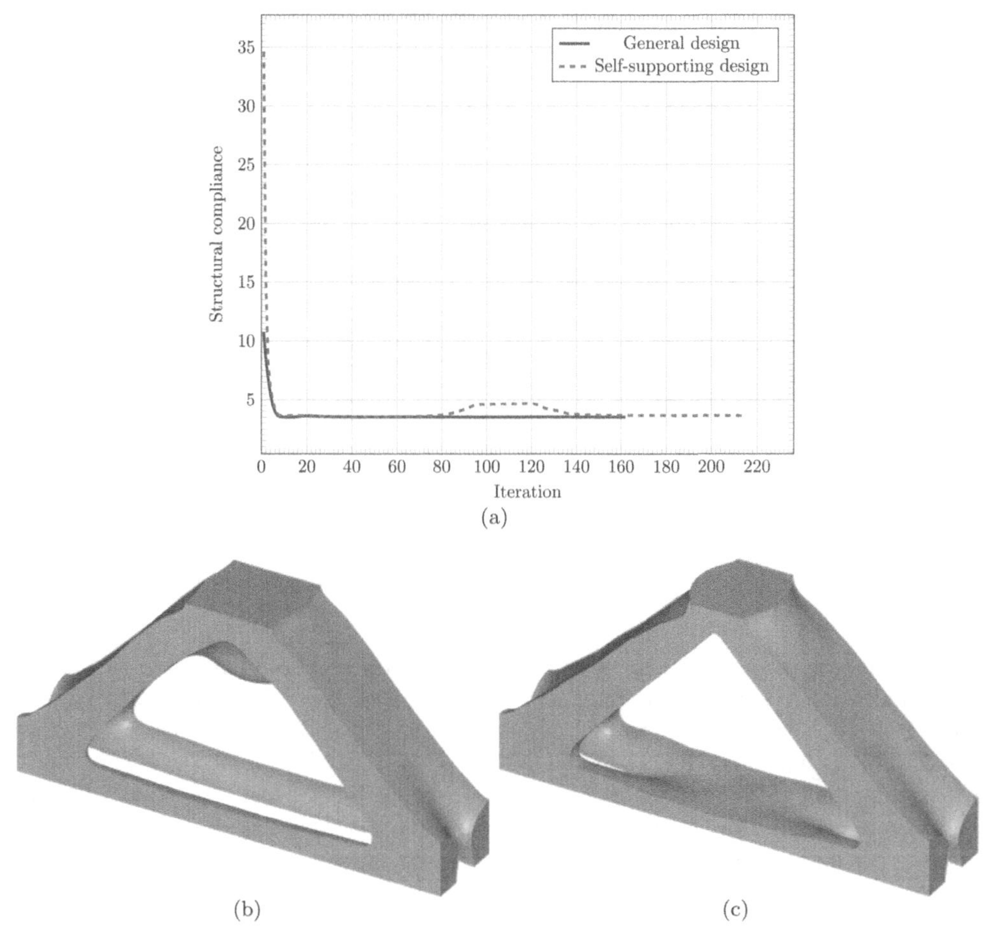

FIGURE 10.5
A simply supported beam subjected to $V^* = 0.3$: (a) optimization processes, (b) general design, and (c) self-supporting design.

TABLE 10.1
Structural compliance and number of iterations of self-supporting structures for a simply supported beam subjected to different target volume fractions

Target volume fraction V^*	**0.2**	**0.4**	**0.5**	**0.6**	**0.7**	**0.8**
Structural compliance	3.982	3.413	3.328	3.292	3.265	3.248
Number of iterations	2,009	162	179	299	205	227

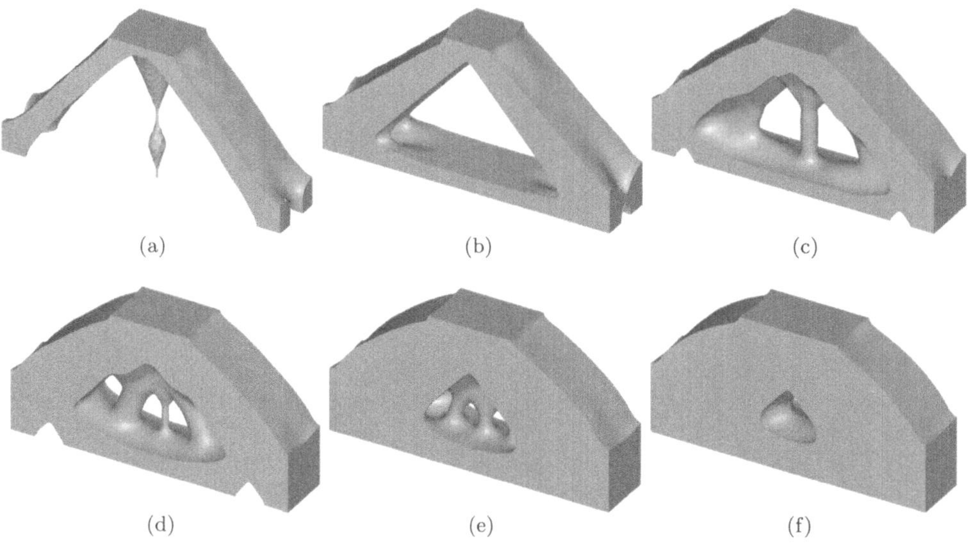

FIGURE 10.6
Optimized topologies of a simply supported beam under different volume fractions: (a) $V^* = 0.2$, (b) $V^* = 0.4$, (c) $V^* = 0.5$, (d) $V^* = 0.6$, (e) $V^* = 0.7$, and (f) $V^* = 0.8$.

eral and self-supporting topology optimization cases are executed subjected to $V^* = 0.15$, and optimization processes and their corresponding topologies are shown in Figure 10.7. Figure 10.7(a) shows that the general design converges after 157 iterations. By contrast, it takes 1,507 iterations for self-supporting design to converge. The objective values of general and self-supporting designs are 4.063 and 4.257, respectively. Most importantly, severe oscillations are observed in the convergence process of self-supporting design. Comparing the structures in Figure 10.7(b) and 10.7(c), despite having severe numerical oscillations, the self-supporting topology still can be successfully formed.

10.4.2 Effects of Varying Mesh Size

Effects of the mesh size are investigated with a long simply supported beam presented in Figure 10.8, where the build orientation is in the $+y$ direction. In this case, a mesh of $120\times20\times20$, $V^* = 0.35$, and $r_{\text{min}} = 3$ are considered. Optimization processes of general and self-supporting topology optimization and their corresponding topologies are shown in Figure 10.9. Figure 10.9(a) shows that both optimization processes converge after around 180 iterations; however, there is a severe numerical oscillation in the convergence process of self-supporting design. The objective values of general design and self-supporting design are 24.821 and 24.459, respectively. Interestingly, the performance of self-supporting design is slightly better than that of general design. It seems that the self-supporting design in Figure 10.9(c) is more reasonable than general design in Figure 10.9(b).

Structural performance and convergence of self-supporting structures for a long simply supported beam under different mesh sizes (that is, $60\times10\times10$, $90\times15\times15$, $180\times30\times30$, and $240\times40\times40$) are outlined in Table 10.2. As the element size is scaled with a certain ratio, the filter radius is also scaled with the same ratio. The finer mesh can always lead to improved results since more design freedom is provided. In addition, there is no clear relationship between the mesh size and number of iterations. The resulting topologies are shown in Figure 10.10, where the structural smoothness increases when using the fine mesh.

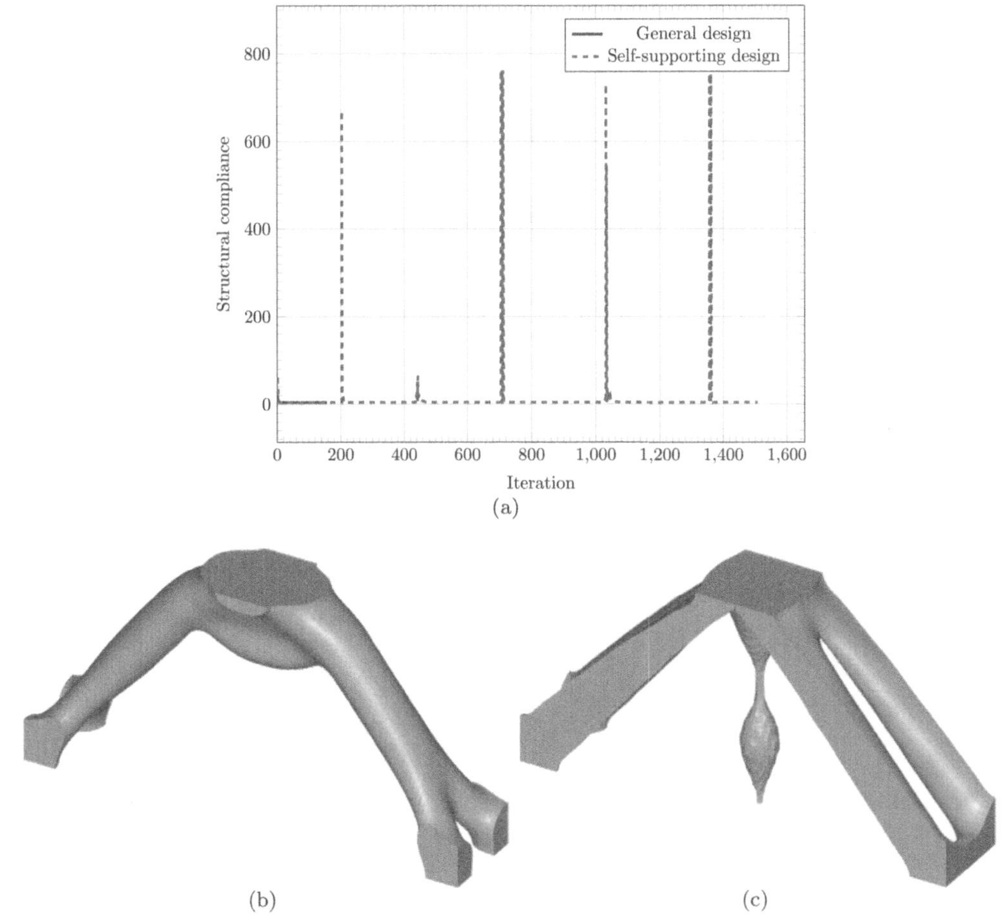

(a)

(b) (c)

FIGURE 10.7
A simply supported beam subjected to $V^* = 0.15$: (a) optimization processes, (b) general design, and (c) self-supporting design.

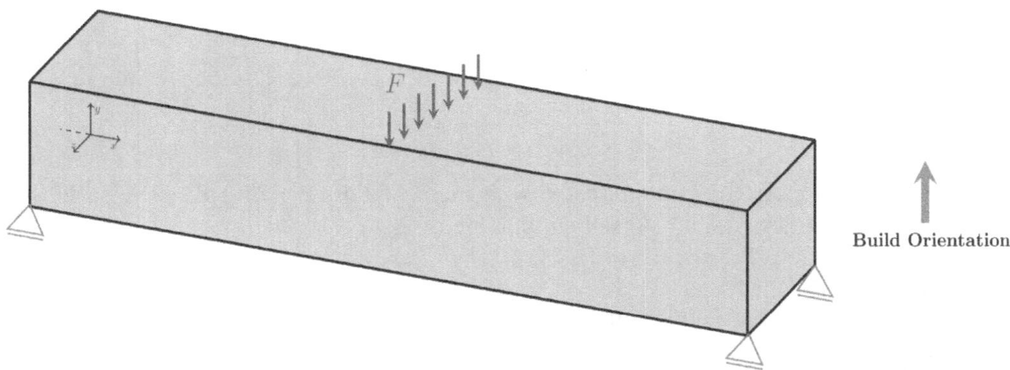

FIGURE 10.8
Design domain of a long simply supported beam.

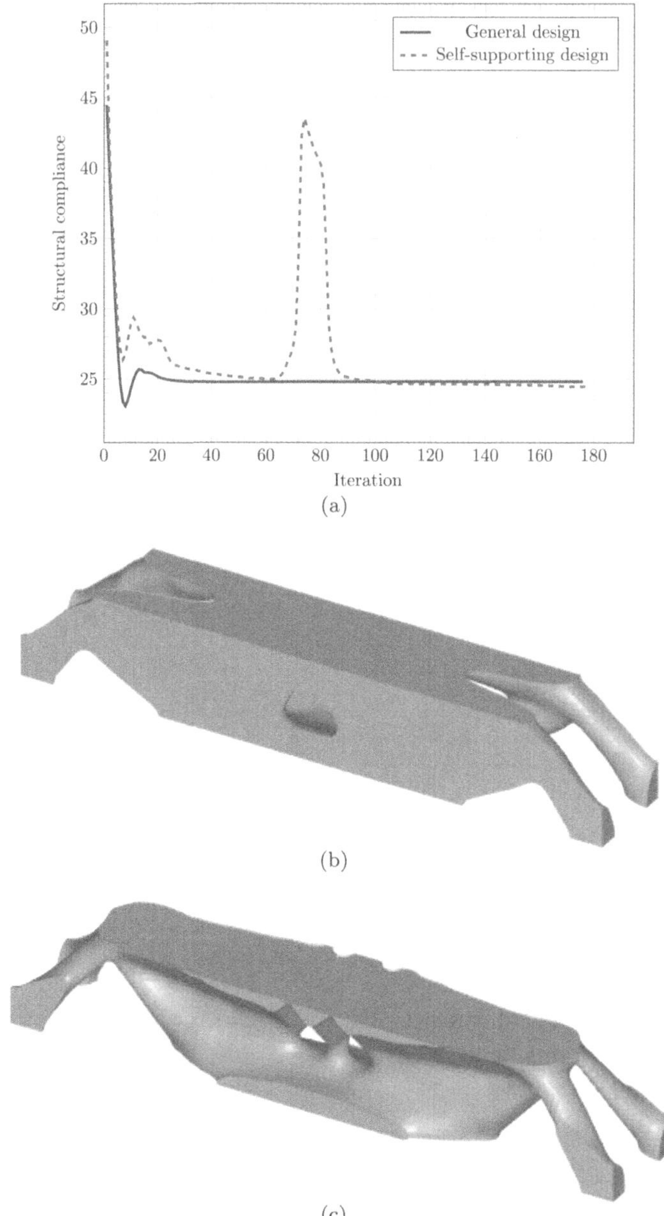

(a)

(b)

(c)

FIGURE 10.9
A long simply supported beam subjected to a mesh of 120×20×20: (a) optimization processes, (b) general design, and (c) self-supporting design.

TABLE 10.2
Structural compliance and number of iterations of self-supporting structures for a long simply supported beam subjected to different mesh sizes

Mesh size	**60×10×10**	**90×15×15**	**180×30×30**	**240×40×40**
Structural compliance	40.011	29.700	19.181	18.841
Number of iterations	153	164	407	243

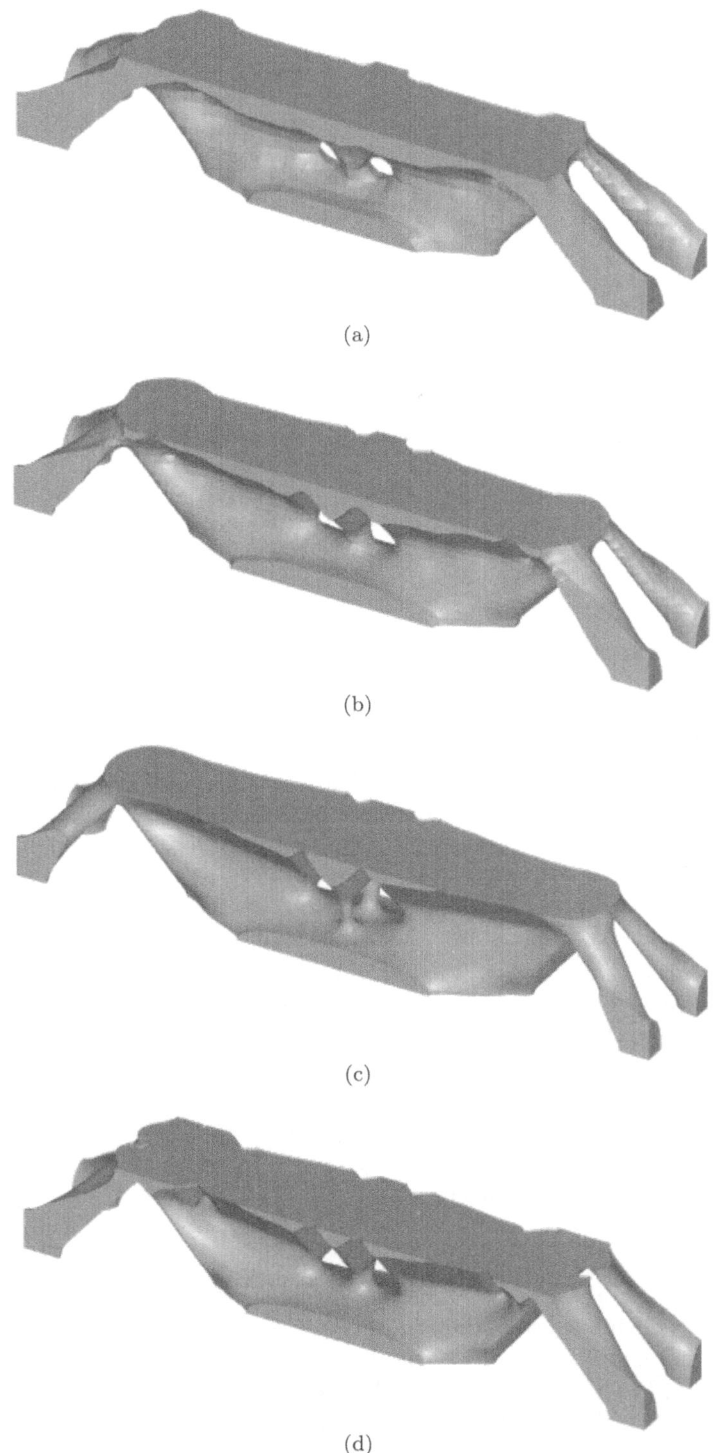

FIGURE 10.10
Resulting topologies of a long simply supported beam under different mesh sizes: (a) 60×10×10, (b) 90×15×15, (c) 180×30×30, and (d) 240×40×40.

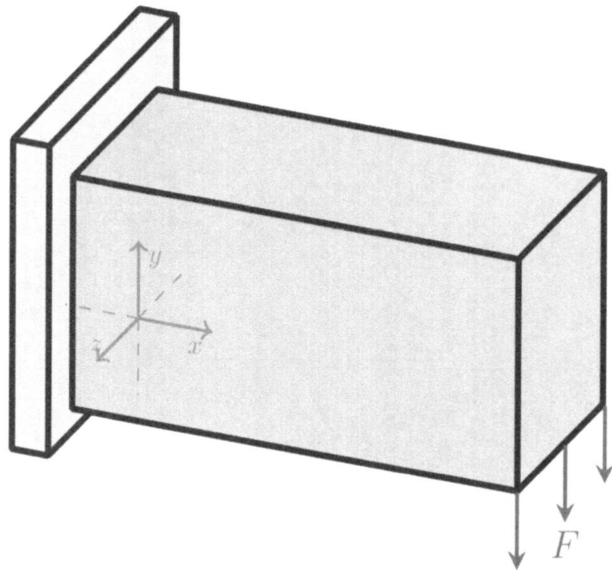

FIGURE 10.11
Cantilever beam with a unit distributed vertical load for self-supporting design.

10.4.3 Effects of Varying Build Orientation

The build direction in AM has significant effect on the formation of self-supporting design. The investigations on the effects of the build direction are carried out using a cantilever beam with a unit distributed vertical load in Figure 10.11, where $+x$ and $+y$ are considered. A mesh of $120\times60\times40$, $V^* = 0.3$, and $r_{\text{min}} = 4$ are adopted in this case. A scaling factor of 0.1 is used for the unit distributed loading.

The convergence processes and corresponding topologies are shown in Figure 10.12. The severe numerical oscillations are observed in the convergence process of the $+y$ direction. The dripping issue can be found in Figure 10.12(d), which is illustrated in Figure 10.13. Although these drop-like features satisfy the critical overhang angle of 45°, they are not printable since there are no sufficient supports [56]. Therefore, the dripping effect should be prevented. In SEMDOT, a straightforward way is reducing the evolution rate. Here, the evolution rate of 0.1 ($\Lambda = 0.1$) is used. The resulting topology with $\Lambda = 0.1$ is shown in Figure 10.12(e), where no dripping issue is observed. In addition, the numerical oscillations of the convergence process are alleviated.

10.4.4 Effects of Varying Parameter for Approximation Smoothness

This case studies the influences of the P-norm parameter or parameter for the approximation smoothness P in the AM filter on final results. The cantilever beam with a unit vertical load shown in Figure 10.14 is used for this case. In addition, a mesh of $120\times60\times40$, $V^* = 0.3$, and $r_{\text{min}} = 4$ are utilized. Optimization processes of general and self-supporting topology optimization and their topologies are shown in Figure 10.15. Optimization processes of general and self-supporting designs converge after around 800 iterations, and their corresponding objective function values are 5.938 and 6.075, respectively. When using $P = 40$, the hole in Figure 10.15(c) is altered to satisfy the critical overhang angle of 45° in comparison with general design (Figure 10.15b).

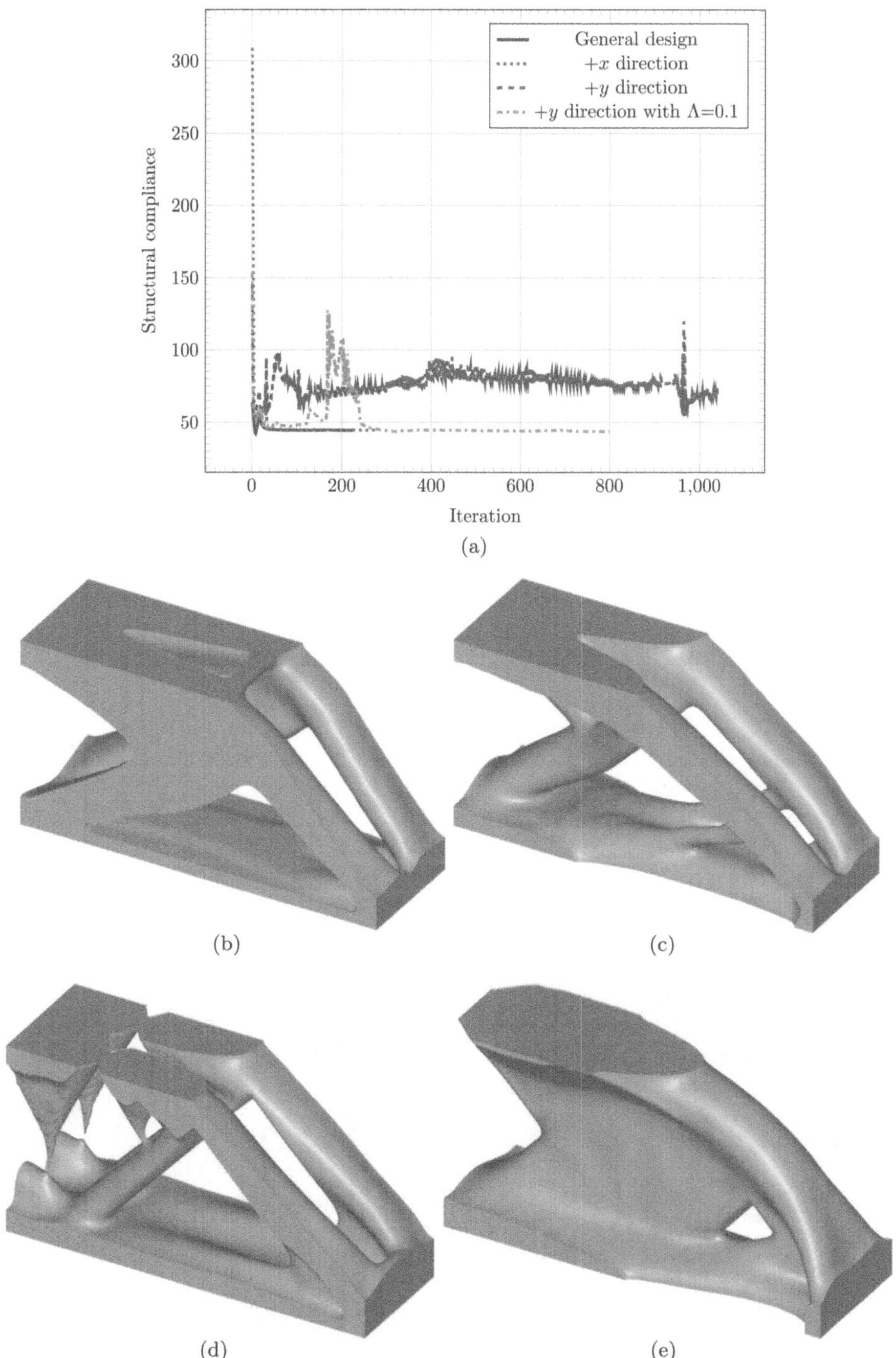

FIGURE 10.12
Resulting topologies of a long simply supported beam under different build orientations: (a) general design, (b) $+x$ direction, (c) $+y$ direction, and (d) $+y$ direction with Λ=0.1.

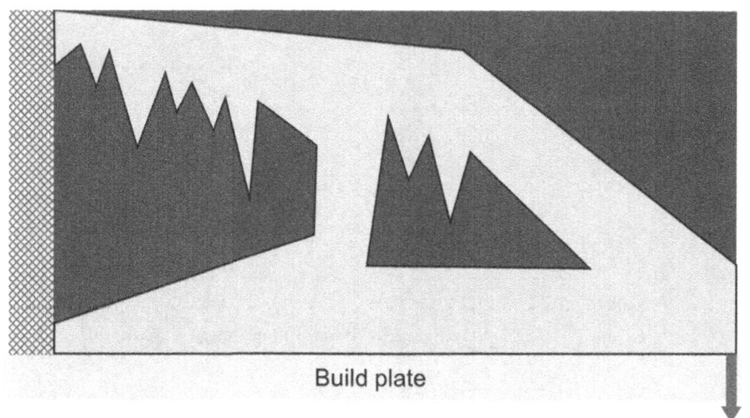

FIGURE 10.13
Illustration of dripping issue in additive manufacturing.

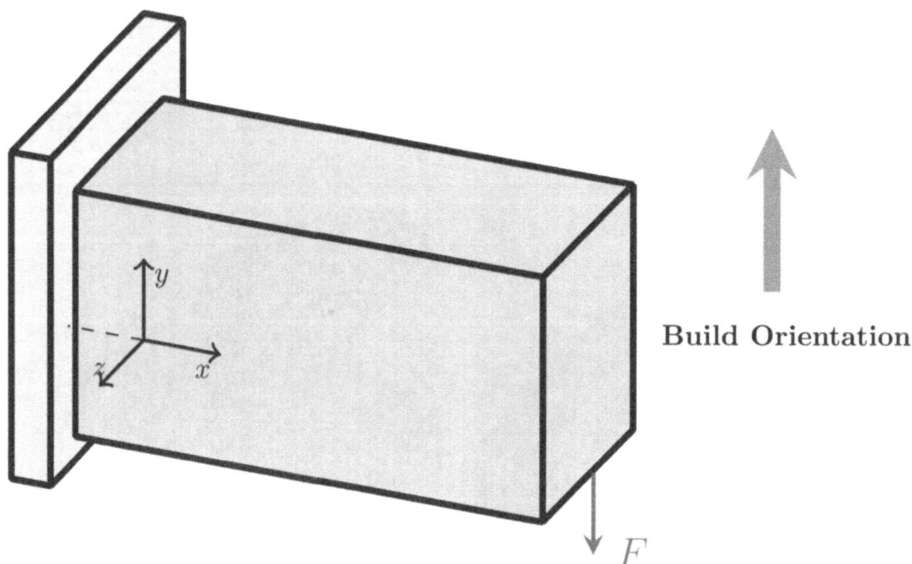

FIGURE 10.14
Cantilever beam with a unit vertical load for self-supporting design.

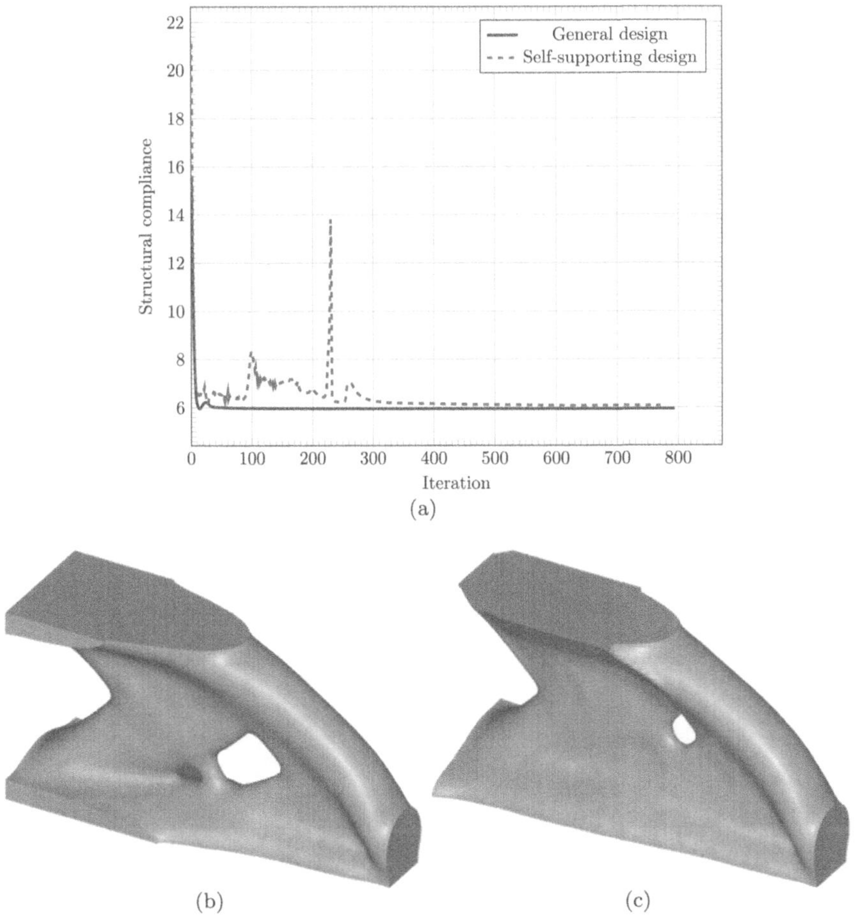

FIGURE 10.15
Cantilever beam with a unit vertical load for self-supporting design: (a) optimization processes, (b) general design, and (c) self-supporting design with $P = 40$.

When considering different approximation smoothness parameters P (P-norm parameters), compliance and number of iterations of self-supporting structures for a cantilever beam with a unit vertical load are summarized in Table 10.3. With the increase in P, the structural performance improves. A small value of P (for example, $P = 10$) results in a long convergence process with a total number of iterations of 932. The corresponding topological configurations are shown in Figure 10.16, where asymmetric designs are observed when using $P = 60$ and $P = 80$.

10.4.5 Large-Scale Self-Supporting Design Case

To further demonstrate the effectiveness of the combination of SEMDOT and Langelaar's AM filter, a large-scale self-supporting design case is taken into account. A mesh of $180 \times 90 \times 64$ (more than one million elements), $V^* = 0.3$, and $r_{\min} = 4$ are used for this large-scale case. In the AM filter, the P-norm parameter P is set to 30. To alleviate the computational cost, the number of grid points of $3 \times 3 \times 3$ in each element is adopted. To stabilize

TABLE 10.3
Structural compliance and number of iterations of self-supporting structures for a cantilever beam with a unit vertical load subjected to different approximation smoothness parameters

Approximation smoothness parameter P	**10**	**30**	**60**	**80**
Structural compliance	6.081	6.182	6.010	5.963
Number of iterations	932	761	765	786

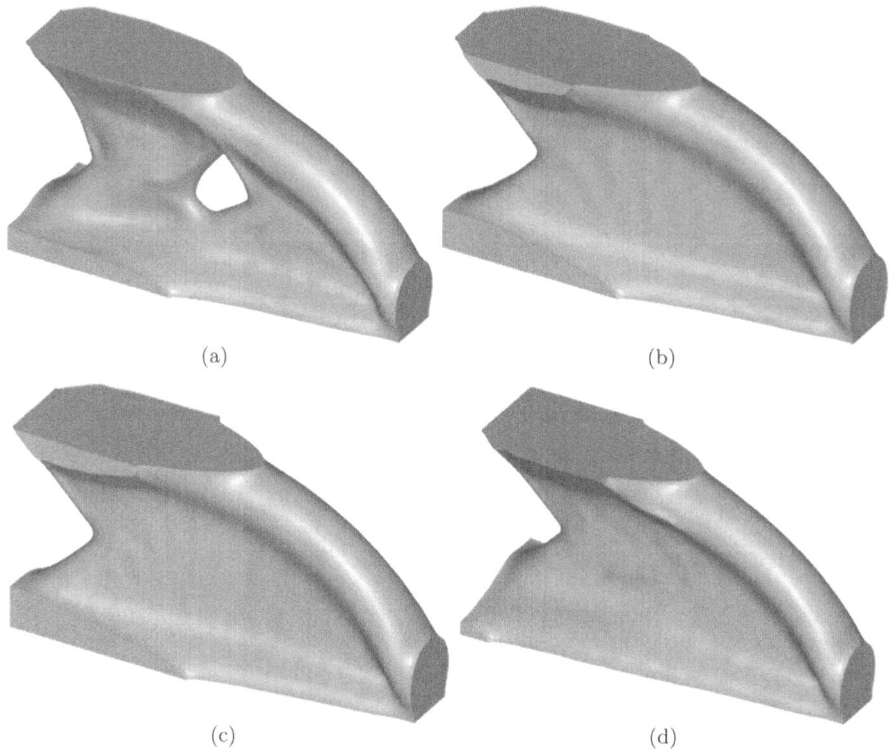

FIGURE 10.16
Resulting topologies of a cantilever beam with a unit vertical load under different P-norm parameters: (a) $P = 10$, (b) $P = 30$, (c) $P = 60$, and (d) $P = 80$.

the convergence process, the evolution rate of 0.1 is used in the Heaviside smooth function. For some optimization cases, the topological boundary error criterion is too conservative, resulting in an unnecessarily long convergence process. Therefore, the tolerance value of the topological boundary error is increased to 0.01 from 0.001. To solve the large-scale self-supporting design problem efficiently, the iterative solver, the Preconditioned Conjugate Gradient (PCG) method [13], is employed instead of the traditional direct solver. The PCG procedure is mathematically described by:

$$\mathbf{u} = PCG(\mathcal{M}^{-1}, \mathbf{K}, \mathbf{F}, \mathbf{u}_0), \tag{10.12}$$

where $\mathbf{u}_0$ is the input vector of nodal displacements, $\mathbf{u}$ is the output vector of nodal displacements, $\mathcal{M}$ is the preconditioner matrix, $\mathbf{K}$ is the global stiffness matrix, and $\mathbf{F}$ is the vector of nodal forces. In PCG, the tolerance is set to 1×10^{-8}, and the maximum number of iterations is set to 8,000.

Figure 10.17(a) shows that the optimization process converges at the structural compliance of 5.551 after 711 iterations. As shown in Figure 10.17(b), the self-supporting design

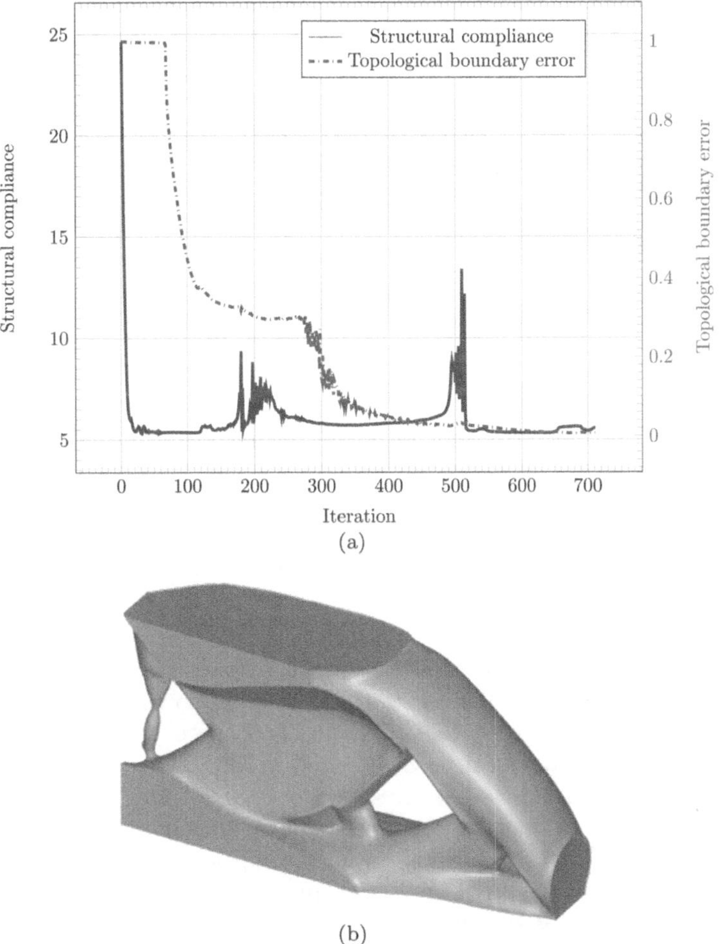

FIGURE 10.17
Large-scale self-supporting design case: (a) convergence process and (b) optimized topology.

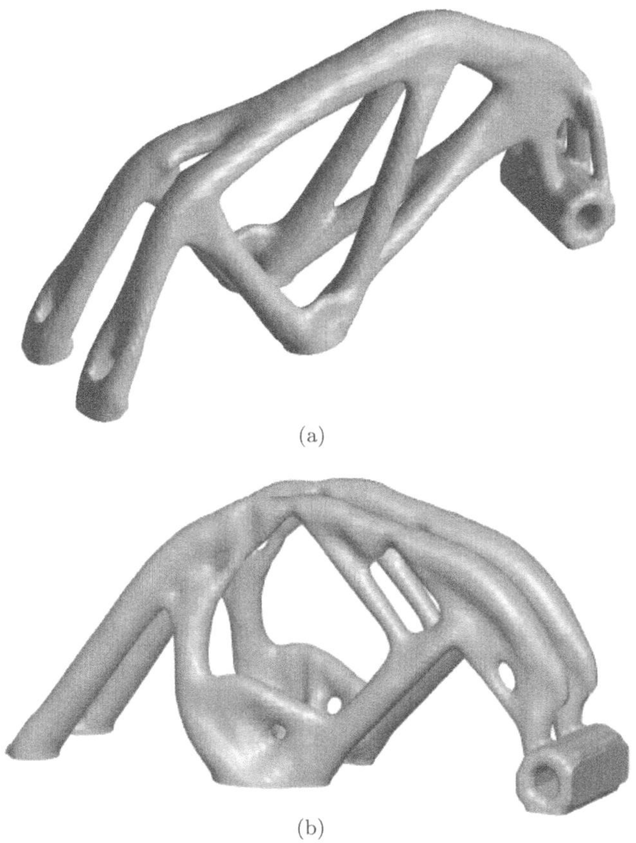

(a)

(b)

FIGURE 10.18
Industrial frame: (a) general design and (b) self-supporting design.

can be successfully generated; however, the optimized topology is not symmetric. The asymmetric topology should not be regarded as an inappropriate design as long as the critical overhang angle of 45° can be satisfied. If the symmetric design is required, half of the design domain can be used.

10.4.6 Industrial Frame Case

Finally, the industrial frame case presented in [21] is used to show the capability of the combination of SEMDOT and Langelaar's AM filter in solving the real-world problem. To reduce the computational cost, half of the design domain is considered here. More details about the industrial frame case can be found in [83]. The general and self-supporting topologies of the industrial frame case are shown in Figure 10.18. A large proportion of the geometry in Figure 10.18(a) violates the overhang angle condition. By contrast, the geometry in Figure 10.18(b) satisfies the overhang angle condition. The optimized self-supporting design is successfully manufactured with both plastic and metallic materials without the need of support materials, as shown in Figure 10.19.

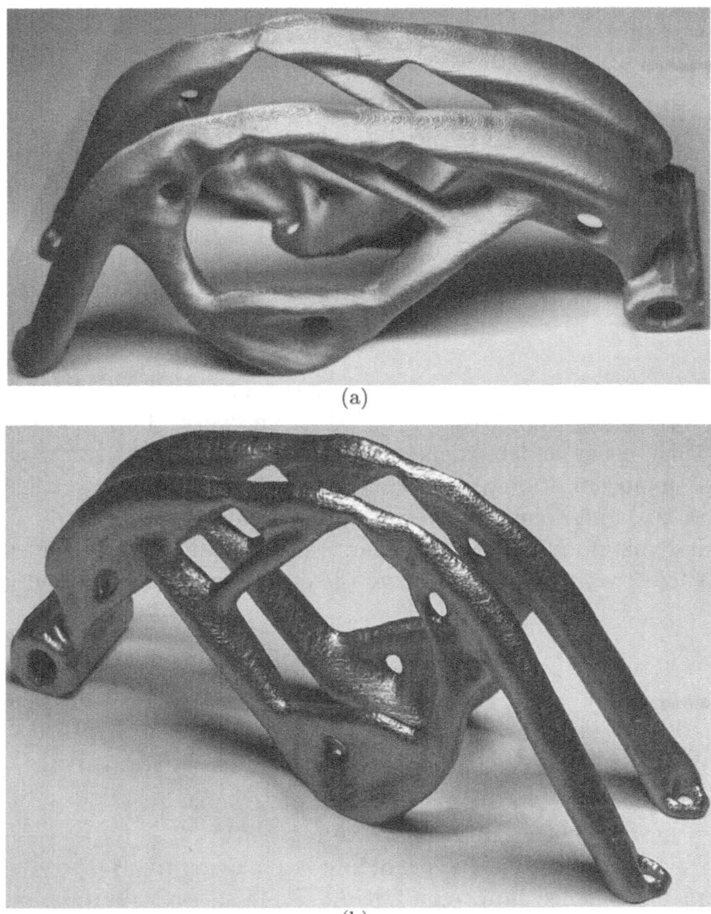

(a)

(b)

FIGURE 10.19
Additively manufactured industrial frame with (a) plastic and (b) metallic materials.

10.5 Summary

This chapter successfully generated 3D self-supporting designs via the combination of SEMDOT and Langelaar's AM filter. The evolution rate of 0.1 in the Heaviside smooth function is recommended for self-supporting design to alleviate severe numerical oscillations and dripping issues. Although asymmetric self-supporting designs are observed, the critical overhang angle of 45° is satisfied. The effectiveness of the proposed self-supporting design methodology is experimentally validated by the industrial framework case.

11

Homogenization Topology Optimization Problems

This chapter incorporates the homogenization theory into the Smooth-Edged Material Distribution for Optimizing Topology (SEMDOT) framework. Two representative homogenization problems, shear and bulk modulus maximization, are used to demonstrate the effectiveness of homogenization SEMDOT via both 2D and 3D cases. Effects of the initial design, filter radius, and volume fraction on final results are discussed. The capability of SEMDOT in designing unit cells with a coarse mesh is validated. In the end, compression testing is conducted to further demonstrate the effectiveness of optimized lattices.

11.1 Optimization Problem

A cellular structure can be regarded as a composite comprising solid and void phases. For the composites having sufficiently regular heterogeneities, it is easy to assume a periodic structure for the composite [71]. The procedure of replacing the composite with an equivalent material model is called **homogenization** [71]. This procedure (homogenization) can overcome the difficulties in analyzing the boundary value problems with a large number of heterogeneities. The homogenization approach is suitable for the composites with periodic micro-structures. The inverse homogenization approach was first used by Sigmund [155] to carry out topology optimization for the material design. Afterwards, the investigations on the bulk modulus [59], effective Young's modulus [114], architected materials [139], and structures with negative, positive, and zero Poisson's ratios [26, 92] were conducted using homogenization topology optimization. Nowadays, the bone scaffold design can exemplify the applications of homogenization topology optimization [119, 132, 165, 187]. Open source codes of homogenization topology optimization can be found in [9, 195].

The contents of this chapter are mainly based on the work by the authors [48], and some improvements and extensions are made in this chapter. Under the condition of linear elasticity, the equivalent constitutive behavior of periodically patterned microstructures in the homogenization method is illustrated in Figure 11.1. To realize homogenization Smooth-Edged Material Distribution for Optimizing Topology (SEMDOT), the homogenization theory presented in [69, 70] is integrated into the SEMDOT framework. Based on [155], the elasticity tensor is given by:

$$E_{ijkl} = \frac{1}{|V^*|}\int_V E_{pqrs}(\varepsilon_{pq}^{0(ij)} - \varepsilon_{pq}^{(ij)})(\varepsilon_{rs}^{0(kl)} - \varepsilon_{rs}^{(kl)})dV^*, \tag{11.1}$$

where V^* is the unit cell volume fraction, E_{pqrs} is the locally varying stiffness tensor, $\varepsilon_{pq}^{0(ij)}$ is the predefined macroscopic strain field, and $\varepsilon_{pq}^{(ij)}$ is the locally varying strain field induced by imposing $\varepsilon_{pq}^{0(ij)}$ on the boundaries of the unit cell.

DOI: 10.1201/9781032634449-11

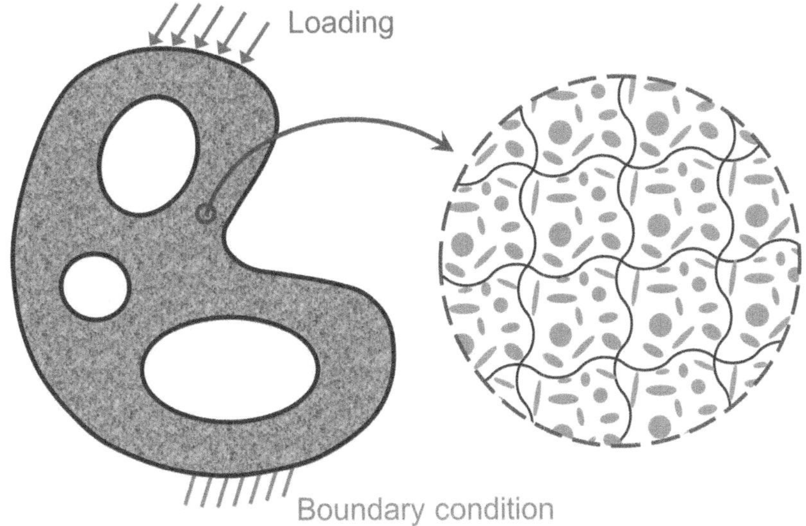

FIGURE 11.1
A material point constituted by periodically patterned microstructures [195].

On account of the discretization in Finite Element Analysis (FEA), the elasticity tensor in homogenization SEMDOT can be rewritten as:

$$E_{ijkl} = \frac{1}{|V^*|} \sum_{e=1}^{M} \left(\mathbf{u}_e^{(ij)}\right)^{\mathrm{T}} \mathbf{k}_1 \mathbf{u}_e^{(kl)}, \tag{11.2}$$

where e is the number of the element, M is the total number of elements in the unit cell, $\mathbf{u}_e^{(kl)}$ is the elemental displacement solution to the unit test strain field, and $\mathbf{k}_1$ is the elemental stiffness matrix.

According to [155], the matrix form of the elasticity tensor for the 2D case is given by:

$$E = \begin{bmatrix} E_{1111} & E_{1122} & E_{1112} \\ E_{1122} & E_{2222} & E_{2212} \\ E_{1112} & E_{2212} & E_{1212} \end{bmatrix}. \tag{11.3}$$

According to [156], the matrix form of the elasticity tensor for the 3D case is:

$$E = \begin{bmatrix} E_{1111} & E_{1122} & E_{1133} & 0 & 0 & 0 \\ E_{2211} & E_{2222} & E_{2233} & 0 & 0 & 0 \\ E_{3311} & E_{3322} & E_{3333} & 0 & 0 & 0 \\ 0 & 0 & 0 & E_{1212} & 0 & 0 \\ 0 & 0 & 0 & 0 & E_{2323} & 0 \\ 0 & 0 & 0 & 0 & 0 & E_{3131} \end{bmatrix}. \tag{11.4}$$

In SEMDOT, the Young's modulus is approximated by a linear interpolation between the two phases of solid and void materials, which has the form:

$$E(X_e) = (1 - X_e)E_{\min} + X_e E_1, \tag{11.5}$$

where E_1 and $E_{\min}$ are elastic moduli of the solid and void materials, respectively.

On the basis of Equations (11.2) and (11.5), the sensitivity of the elasticity tensor in homogenization SEMDOT is calculated by:

$$\frac{\partial E_{ijkl}}{\partial X_e} = \frac{1}{|V*|}((1 - X_e)E_{\min} + X_e E_1)\left(\mathbf{u}_e^{(ij)}\right)^{\mathrm{T}} \mathbf{k}_1 \mathbf{u}_e^{(kl)}. \tag{11.6}$$

In homogenization SEMDOT, periodic boundary conditions and square symmetry are taken into account, and hence we have the following relationships: $E_{1111} = E_{2222}$ and $E_{1122} = E_{2211}$ for 2D cases, and $E_{1111} = E_{2222} = E_{3333}$, $E_{1122} = E_{2211} = E_{2233} = E_{3322} = E_{3311} = E_{1133}$, and $E_{2323} = E_{3131} = E_{1212}$ for 3D cases.

Two representative homogenization topology optimization problems, shear and bulk modulus maximization, are taken into account in this chapter. The optimization problem is therefore mathematically described by:

$$\begin{aligned} &\max : C(X_e) = G \text{ or } K \\ &\text{subject to} : \frac{\sum_{e=1}^{M} X_e V_e}{\sum_{e=1}^{M} V_e} - V^* \leq 0, \\ &0 < \rho_{\min} \leq X_e \leq 1; \; e = 1, 2, \cdots, M, \end{aligned} \tag{11.7}$$

where $C(X_e)$ is the objective function, X_e is the elemental volume fraction, G is the shear modulus, K is the bulk modulus, V_e is the volume of the eth element, and $\rho_{\min}$ is a small value, which is 0.001 here.

For 2D cases, the shear modulus is calculated by:

$$G = E_{1212}. \tag{11.8}$$

In 3D cases, the shear modulus is:

$$G = \frac{1}{3}(E_{2323} + E_{3131} + E_{1212}). \tag{11.9}$$

For 2D cases, the bulk modulus is calculated by:

$$K = \frac{1}{4}(E_{1111} + E_{1122} + E_{2211} + E_{2222}). \tag{11.10}$$

In 3D cases, the bulk modulus is:

$$K = \frac{1}{9}(E_{1111} + E_{1122} + E_{1133} + E_{2211} + E_{2222} + E_{2233} + E_{3311} + E_{3322} + E_{3333}). \tag{11.11}$$

It is noted that, although this chapter merely focuses on shear and bulk modulus maximization problems, homogenization SEMDOT can be easily extended to solve other optimization problems such as effective Young's modulus maximization and negative, positive, and zero Poisson's ratio problems.

11.2 Numerical Examples

11.2.1 Shear Modulus Maximization

Four different initial designs in Figure 11.2 are used to investigate the effects on the final results of 2D shear modulus maximization cases. A mesh of 100×100, filter radius of 4

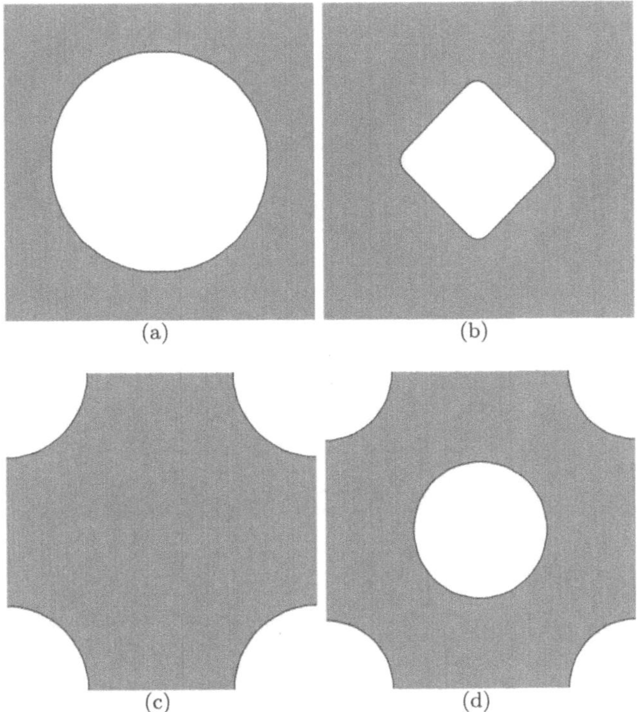

FIGURE 11.2
2D homogenization topology optimization: (a) initial design 1, (b) initial design 2, (c) initial design 3, and (d) initial design 4.

($r_{\min} = 4$), and target volume fraction of 0.35 ($V^* = 0.35$) are utilized. As shown in Figure 11.3, initial designs 1 and 3 lead to the exactly same results. By contrast, initial designs 2 and 4 lead to the same topological configurations; however, the optimized shear modulus values are slightly different.

Two initial designs in Figure 11.4 are used to study the influences on the final results of 3D shear modulus maximization cases. In this case, a mesh of 40×40×40, $r_{\min} = 2$, and $V^* = 0.3$ are adopted. As shown in Figure 11.5, initial designs 1 and 2 lead to two different unit cells, with the corresponding objective function values of 0.067 and 0.0672, respectively. Effective Young's modulus surfaces of the two unit cells for initial designs 1 and 2 are shown in Figure 11.6, where the same distribution is observed, and the high stiffness is along diagonal directions.

The initial design 2 in Figure 11.2(b) is employed to study the effects of the filter radius for 2D cases. The resulting unit cells under a range of filter radii (that is, $r_{\min} = 1.2$, 2, 3, 5, 6, and 7) are shown in Figure 11.7. Different filter radii can lead to distinct topological configurations of the unit cell. When using $r_{\min} = 5$, the shear modulus value is the highest, which is 0.0947.

The initial design 1 in Figure 11.4(a) is utilized to investigate the effects of the filter radius for 3D cases. The shear modulus values of the 3D case under different filter radii are summarized in Table 11.1. Generally, with the increase in the filter radius, the shear modulus value decreases. The corresponding unit cells are shown in Figure 11.8. Different filter radii lead to distinct topological configurations of the unit cell. When the filter radius

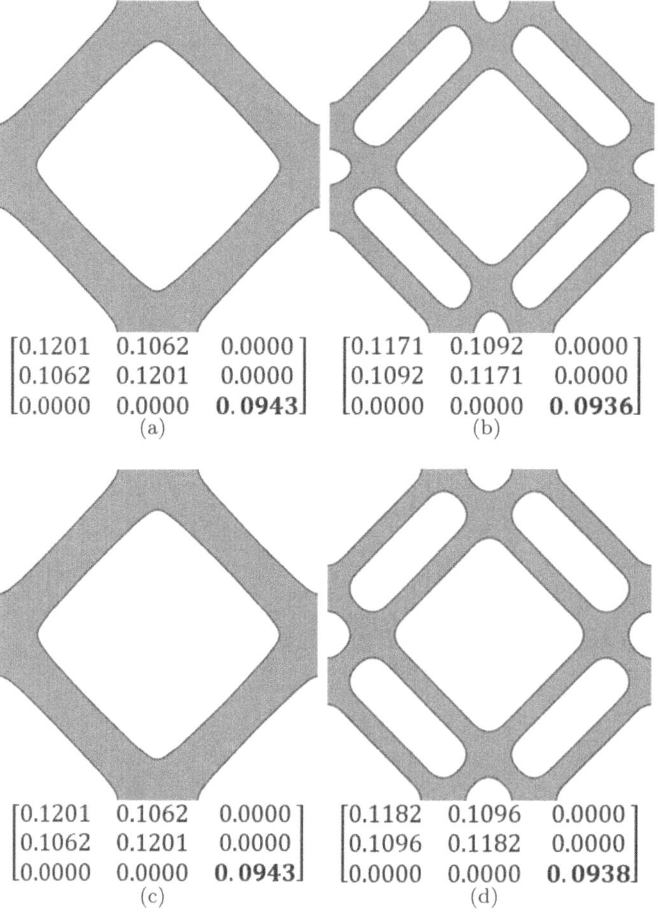

FIGURE 11.3
2D resulting unit cells of shear modulus maximization case for (a) initial design 1, (b) initial design 2, (c) initial design 3, and (d) initial design 4.

is not properly selected (for example, $r_{\min} = 3$), the inappropriate unit cell is obtained, as shown in Figure 11.8(c).

11.2.2 Bulk Modulus Maximization

The Hashin-Strickman (HS) bound is usually used to predict upper and lower limits for certain properties of materials. Here, the HS bound is used to verify the effectiveness of the homogenization SEMDOT algorithm. According to [78, 173], the HS upper bound of the bulk modulus is defined by:

$$K_{\text{upper}} = \frac{V^* K_{HS} G_{HS}}{(1 - V^*) K_{HS} + G_{HS}}, \tag{11.12}$$

where the bulk modulus K_{HS} and shear modulus G_{HS} for the HS upper bound are calculated by:

$$K_{HS} = \frac{E}{3(1 - 2\nu)}, \quad G_{HS} = \frac{E}{2(1 + \nu)}, \tag{11.13}$$

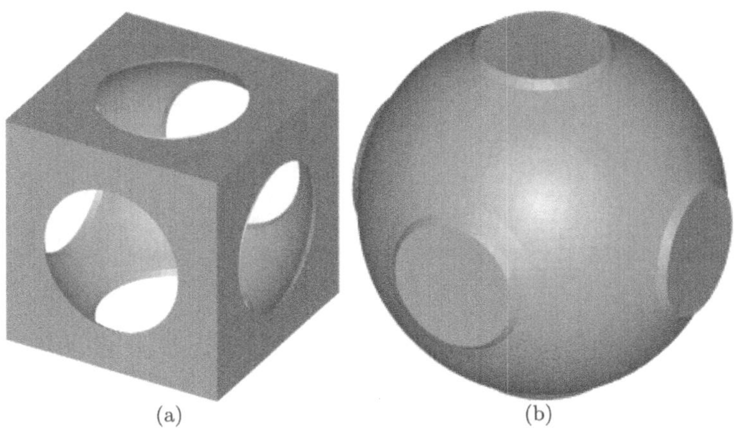

FIGURE 11.4
3D homogenization topology optimization: (a) initial design 1 and (b) initial design 2.

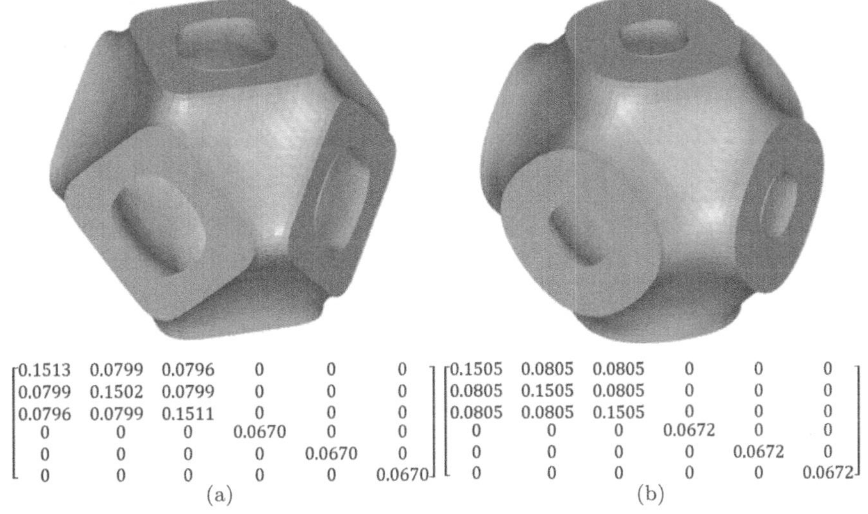

FIGURE 11.5
3D resulting unit cells of shear modulus maximization case for (a) initial design 1 and (b) initial design 2.

TABLE 11.1
Shear modulus of 3D case under different filter radii

Filter radius r_{min}	**1.5**	**2.5**	**3**	**3.5**	**4**	**4.5**
Shear modulus	0.0672	0.0662	0.0582	0.0615	0.0608	0.0575

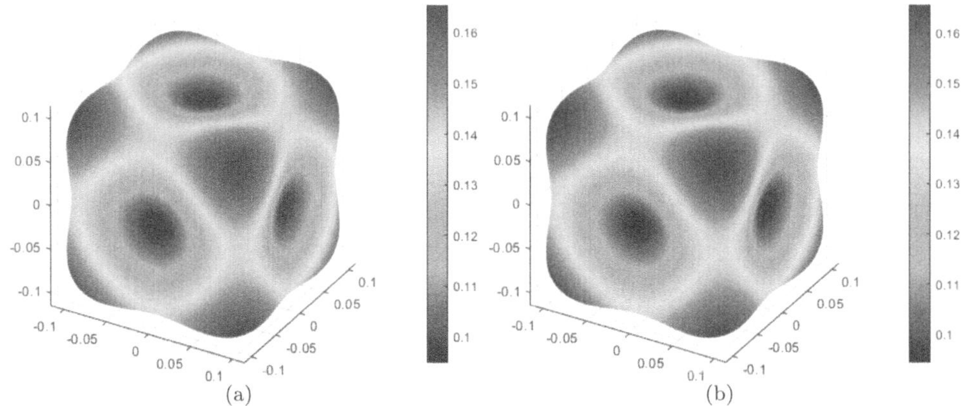

FIGURE 11.6
Effective Young's moduli of shear modulus maximization case for (a) initial design 1 and (b) initial design 2.

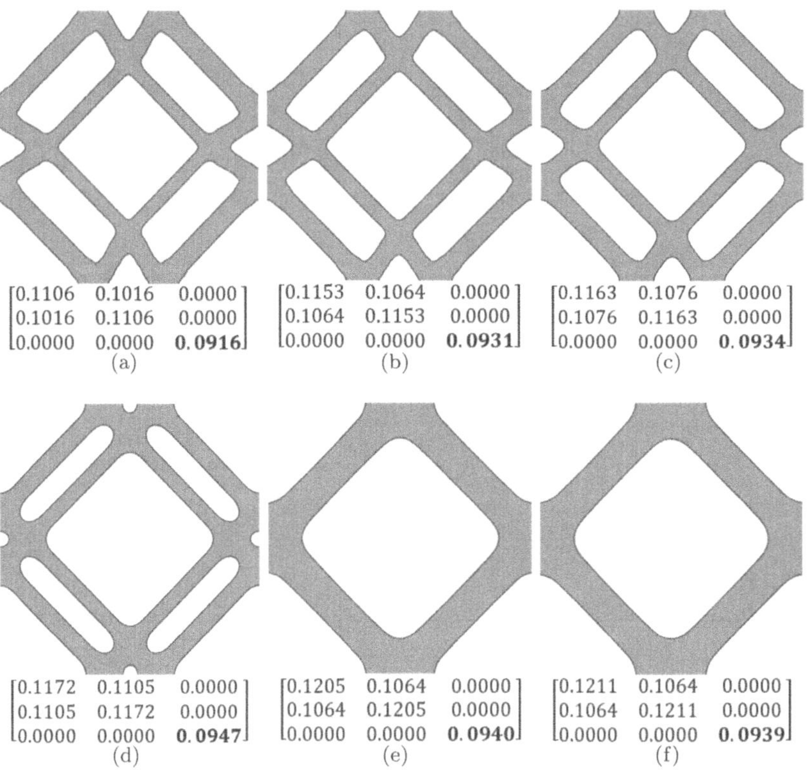

FIGURE 11.7
2D resulting unit cells of shear modulus maximization case under different filter radii: (a) $r_{\min} = 1.2$, (b) $r_{\min} = 2$, (c) $r_{\min} = 3$, (d) $r_{\min} = 5$, (e) $r_{\min} = 6$, and (f) $r_{\min} = 7$.

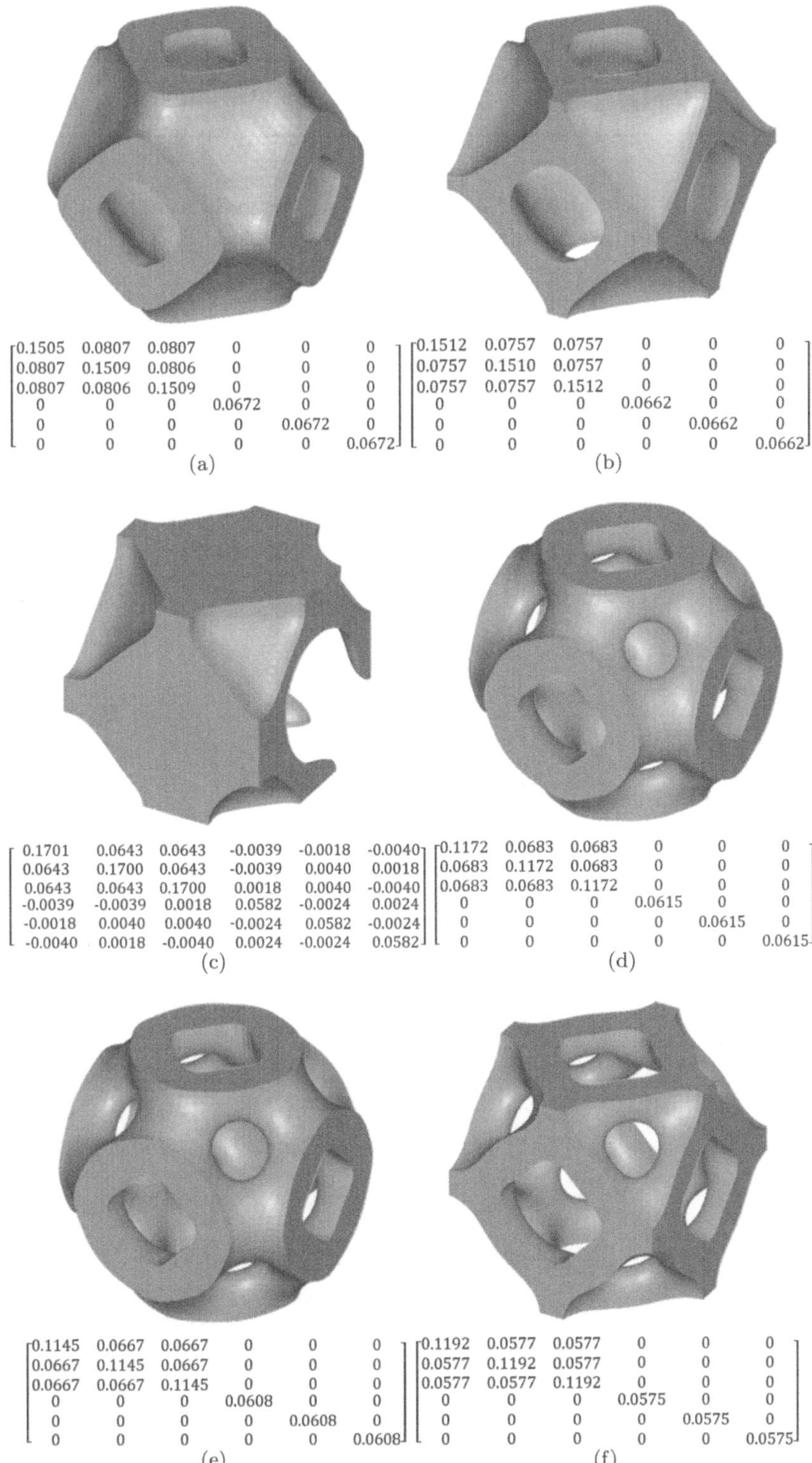

FIGURE 11.8
3D resulting unit cells of shear modulus maximization case under different filter radii: (a) $r_{\min} = 1.5$, (b) $r_{\min} = 2.5$, (c) $r_{\min} = 3$, (d) $r_{\min} = 3.5$, (e) $r_{\min} = 4$, and (f) $r_{\min} = 4.5$.

TABLE 11.2
Bulk modulus of 2D case under different volume fractions

Volume fraction V^*	0.1	0.2	0.3	0.4	0.55	0.65
Bulk modulus	0.0267	0.0580	0.0936	0.1355	0.2148	0.2817

where E is the Young's modulus and ν is the Poisson's ratio. It is noted that the HS bound is not limited to the form in Equation (11.12) [89], and Equation (11.12) is more suitable for 2D cases rather than 3D cases.

The initial design 1 in Figure 11.2(b) is employed to study the effects of the volume fraction for the 2D bulk modulus maximization case. In addition, a mesh of 100×100, $r_{\min} = 4$, and $V^* = 0.35$ are used. The bulk modulus values under different volume fractions (that is, $V^* - 0.1$, 0.2, 0.3, 0.4, 0.55, and 0.65) are summarized in Table 11.2. With the increase in the volume fraction, the bulk modulus rises. The resulting unit cells of the 2D bulk modulus maximization case under different volume fractions are shown in Figure 11.9. Figure 11.10 indicates that the SEMDOT results are close to the HS upper bound.

When using the initial design 2 presented in Figure 11.4(b), an example of the 3D bulk modulus maximization case, which has a mesh of 40×40×40, $r_{\min} = 2$, and $V^* = 0.3$, is

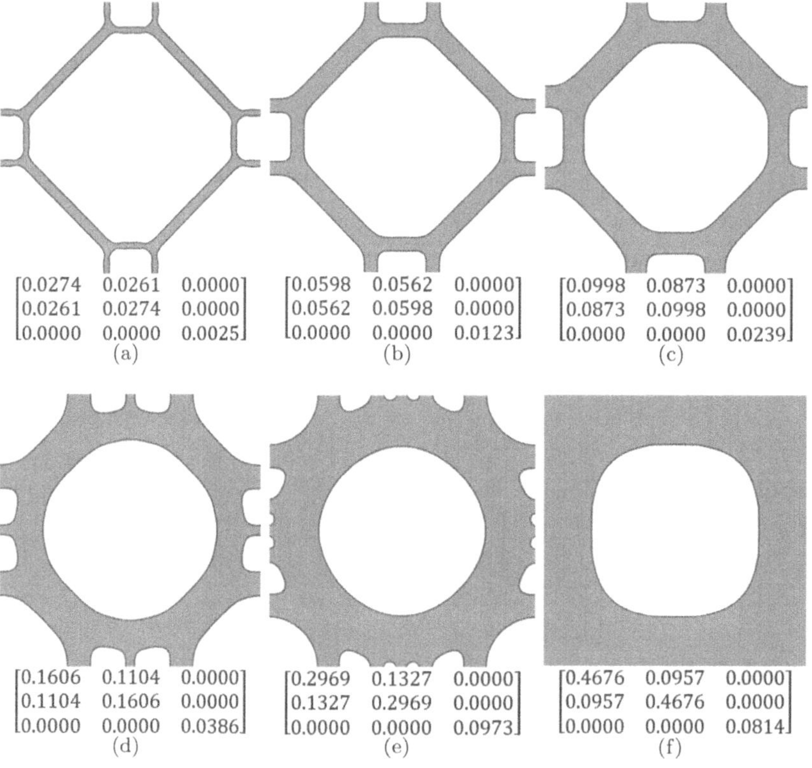

FIGURE 11.9
2D resulting unit cells of bulk modulus maximization case under different volume fractions: (a) $V^* = 0.1$, (b) $V^* = 0.2$, (c) $V^* = 0.3$, (d) $V^* = 0.4$, (e) $V^* = 0.55$, and (f) $V^* = 0.65$.

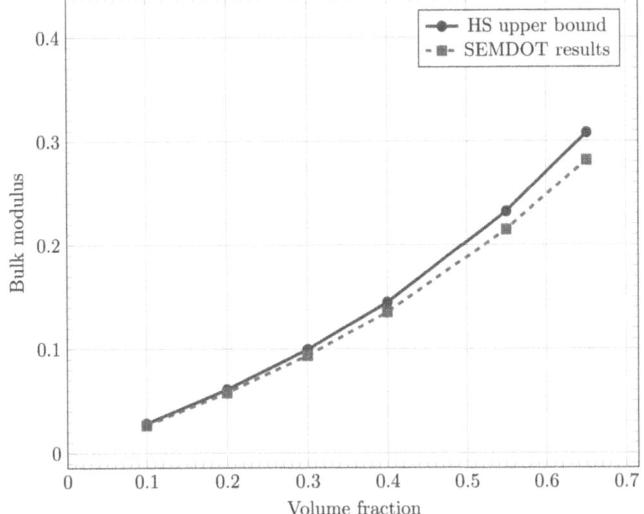

FIGURE 11.10
Comparison between HS upper bound and SEMDOT results for 2D bulk modulus maximization case.

illustrated in Figure 11.11. Figure 11.11(a) shows that the optimization process converges at the bulk modulus of 0.11861 after 227 iterations. The resulting unit cell is shown in Figure 11.11(b), and the corresponding effective Young's modulus surface is shown in Figure 11.11(c), where the high stiffness is along axial directions.

11.3 Unit Cell Design with a Coarse Mesh

An obvious advantage of SEMDOT is that the unit cell can be formed with a very coarse mesh. A 3D case with a coarse mesh of 15×15×15 is taken as an example here. In addition, $r_{\min} = 1.5$ and $V^* = 0.3$ are used. The resulting unit cells for shear and bulk modulus maximization cases are shown in Figure 11.12. The objective function values of shear and bulk modulus maximization cases are 0.0606 and 0.0875, respectively. Although the coarse mesh is used, the topological boundaries presented in Figure 11.12 are reasonably smooth. The corresponding effective Young's modulus surfaces are shown in Figure 11.13.

11.4 Compression Testing

The compression testing is used to further demonstrate the effectiveness of optimized unit cells. A 2D unit cell with the objective of bulk modulus maximization is taken as an example. The MATLAB-based software package proposed by Vogiatzis et al. [176] is used to generate the STereoLithography (STL) file for 3D printing. The parameter setting to generate the STL file is shown in Figure 11.14 where a 4×4 lattice is generated.

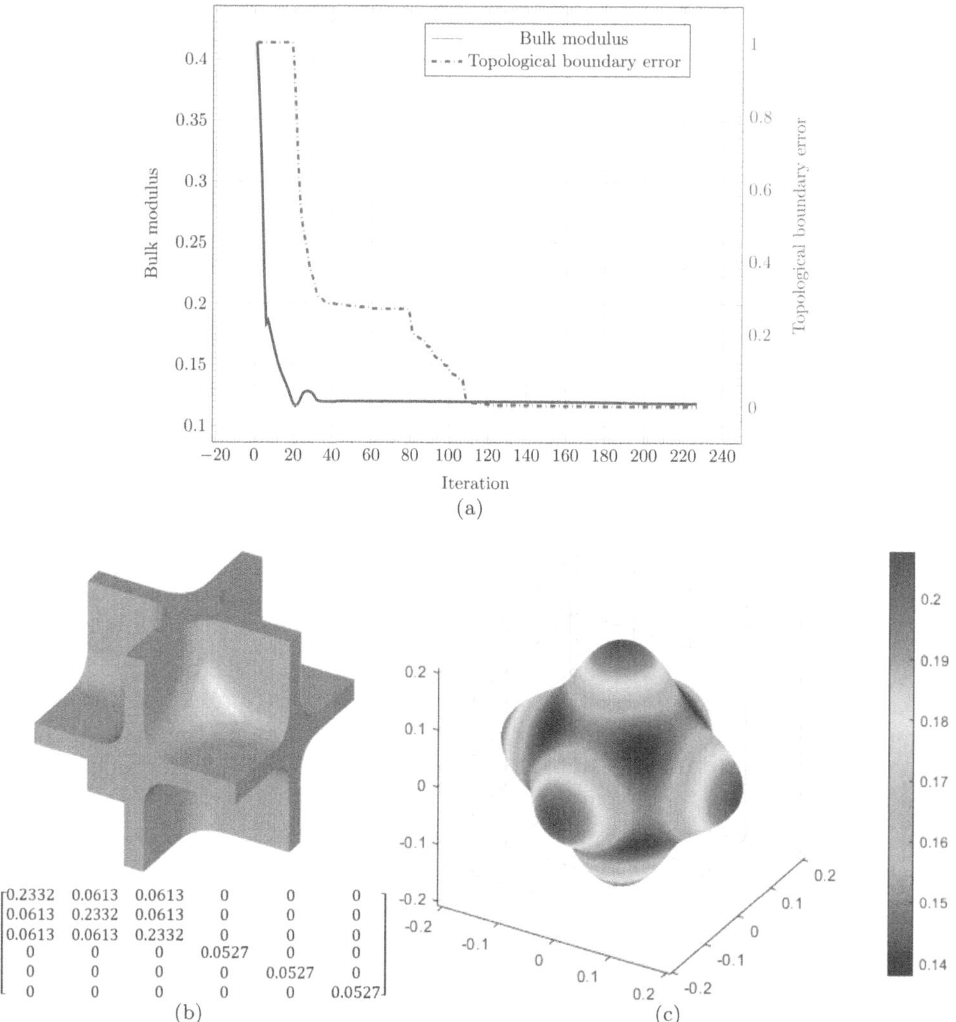

FIGURE 11.11
An example of 3D bulk modulus maximization: (a) optimization process, (b) resulting unit cell, and (c) effective Young's modulus surface.

The latticed structure before and after the compressive load are shown in Figure 11.15. As shown in Figure 11.15(b), the damage is caused by the shear stress. This is because the effective Young's modulus of the lattice is high in axial directions but weak in diagonal directions. Therefore, it can be concluded that the compressive test demonstrates the effectiveness of homogenization SEMDOT. The force versus displacement curve of the 4×4 lattice structure is shown in Figure 11.16.

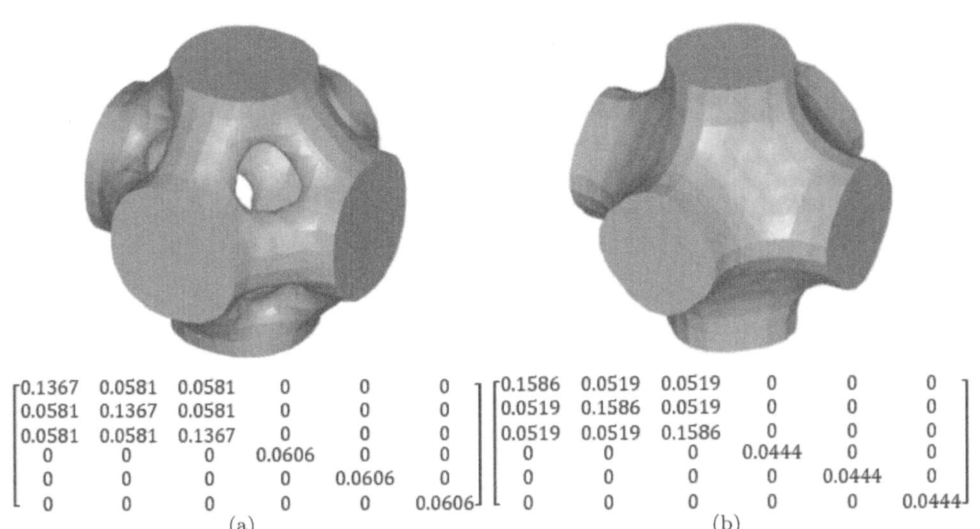

FIGURE 11.12
Unit cell design with a coarse mesh of $15 \times 15 \times 15$: (a) shear and (b) bulk modulus maximization cases.

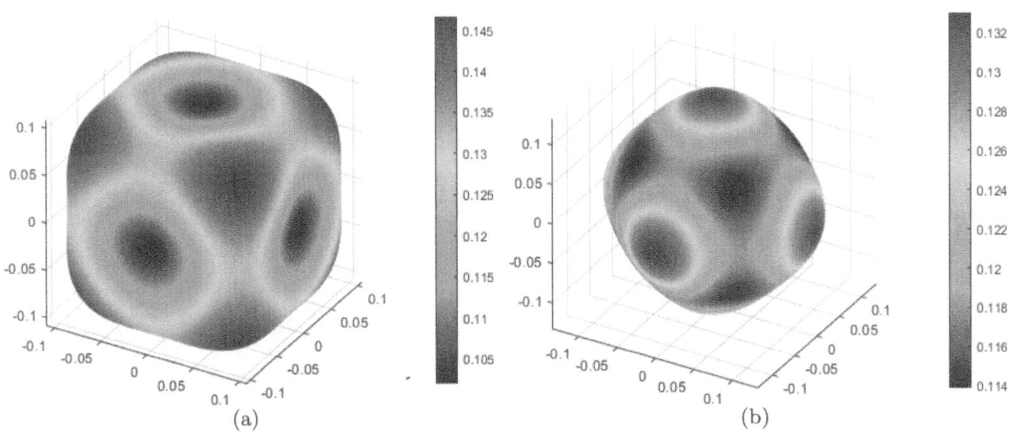

FIGURE 11.13
Effective Young's modulus surfaces with a coarse mesh of 15×15×15: (a) shear and (b) bulk modulus maximization cases.

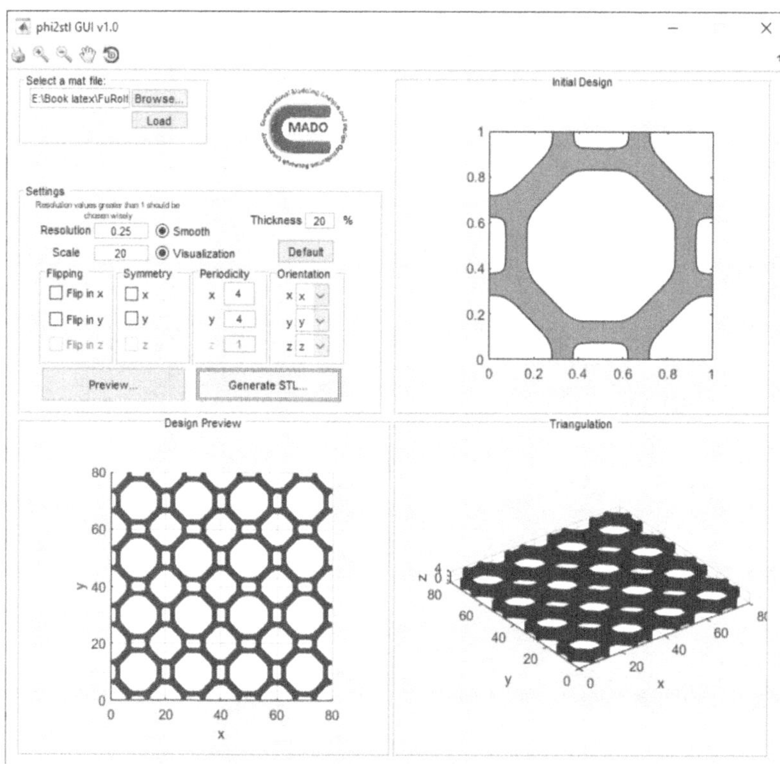

FIGURE 11.14
Parameter setting to generate STL file for lattice.

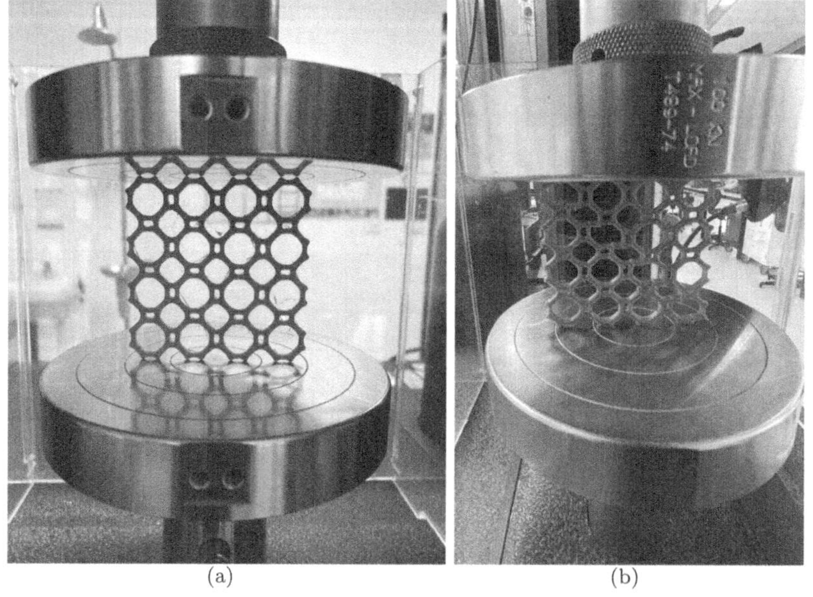

(a) (b)

FIGURE 11.15
Latticed structure (a) before and (b) after compression testing.

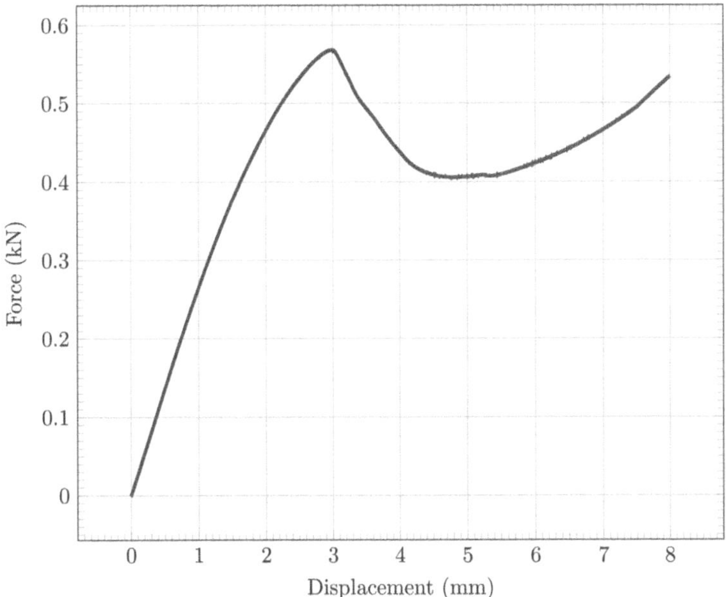

FIGURE 11.16
Force versus displacement curve of latticed structure subjected to compressive loading.

11.5 Summary

The homogenization theory is incorporated into SEMDOT to design the unit cells for shear and bulk modulus maximization. Different filter radii can lead to distinct topological configurations for the unit cell. The improper selection of the filter radius may result in unacceptable unit cells. The 2D bulk maximization case can match the HS upper bound well. SEMDOT can design unit cells with a coarse mesh. The compression testing further demonstrated the effectiveness of optimized unit cells by homogenization SEMDOT.

12

Closing Remarks

This book has been the culmination of many years work on developing the Smooth-Edged Material Distribution for Optimizing Topology (SEMDOT) algorithm. The algorithm started from the development of an idea for reducing the amount of support structures for 3D printed parts. The authors realized with their collaborators that smooth optimized parts are necessary for 3D printing. The SEMDOT algorithm grew from this work. Topology optimization, in its most abstract form, is an algorithm to define paths between loads and constraints within a design volume. The topology optimization process then ranks the significance of material in the design space to then define the "useful" material. As stated in our first chapter, the challenge then is to separate the design volume into solids and voids. The classic SIMP topology optimizaiton algorithm needs a penalization variable, p, to artificially push each element toward either a solid or a void. This leaves a stepped, discontinuous boundary of the final solid(s) based on the finite element mesh resolution and provides complexity when considering multiple materials in the design volume. However, the non-penalization SEMDOT algorithm makes use of an underlying smooth boundary model that drives the "magic" linear interpolation form, see Figure 12.1. This "magic" form simplifies the sensitivity analysis, allowing boundary elements to be weighted sums of solid and void elemental properties. The SEMDOT algorithm reduces the smooth boundary perimeter during the optimization, which helps to separate solid and void elements. Figure 12.1 enables the sensitivity analysis to be based on the separated stiffness matrices of the solids and voids. This coupling of smooth boundaries with finite element analysis using the "magic" linear interpolation form provides a new way to explore different topological designs. The authors believe that there is real potential for the SEMDOT algorithm to be applied to complex multi-phase fluid or material problems. Moreover, the SEMDOT algorithm provides a beneficial topology optimization process for redesigning parts for 3D printing.

This book first described the classical topology optimization techniques that are mainly used in the field (density-based algorithms (SIMP, BESO) and boundary-based algorithms (LSMs, MMC)). The weaknesses of these methods are the lack of simple algorithms that use and produce smooth boundaries. The third chapter described the basic mechanics of the SEMDOT algorithm, and the sensitivity analysis is clearly outlined. This chapter also shows how the SEMDOT algorithm can be used without a penalization approach. Again, the removal of the penalization approach will open up SEMDOT to many applications. Other than SEMDOT, there are other non-penalization algorithms, such as Floating Projection Topology Optimization (FPTO) proposed by Huang [73, 74], who is a pioneer of the non-penalization approach. The authors believe that the current version of SEMDOT is a nice trade-off between the effectiveness of the algorithm and the complexity of the algorithm.

Furthermore, this book is written to enable readers to use the SEMDOT algorithm. The fourth chapter provided information on potential optimizers that can be used to solve topology optimization problems using SEMDOT. The remaining chapters supplied a series of applications where the SEMDOT algorithm has been employed. The first application tackled was the classic stiffness maximization problem for a given mass. The second application

DOI: 10.1201/9781032634449-12

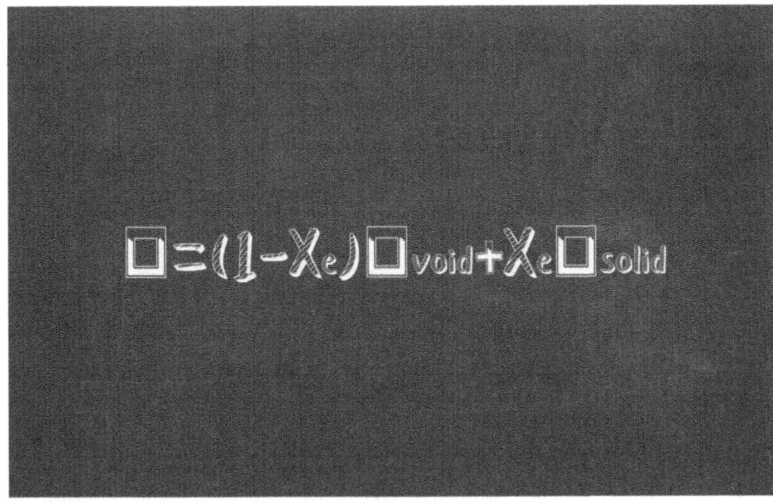

FIGURE 12.1
Magic linear interpolation in SEMDOT.

chapter looked at compliant mechanisms. The third application chapter investigated heat conduction problems. This chapter was split into two parts: the first part examined steady-state heat conduction problems; and the second part investigated the difficult problem of optimizing transient heat conduction. The fourth application chapter examined dynamic problems, including optimizing the natural frequency of a structure, and optimizing the response of a structure given a driving frequency. Advanced topology optimization topics on maximum stress minimization problems, self-supporting designs for additive manufacturing, and homogenization topology optimization problems were also covered in Chapters 9, 10, and 11. Of these topics, SEMDOT has a particular capability for supporting new designs for additive manufacturing with little or no support structures.

Currently, SEMDOT is being considered for use in many other applications, and the readers may want to look through the following literature: multi-material problems [99,143, 202,223], multi-scale problems [63,190,200], fatigue resistance problems [28,29], design for fiber reinforced structures [34,103,201], fluid-based problems [3,5], buckling problems [32,41, 42,203], large-scale problems [1,133,184], gradient-free topology optimization [22,88,90,181], and AI-based approaches [2,12,16,189].

The software to run the problems described in the book are found in Appendix C. As suggested by Sigmund [161], the authors provided all parameters, model, and optimization settings, as well as source codes, for the purpose of reproducibility. The authors will endeavor in the future to simplify the SEMDOT algorithm by reducing the input parameters, and they aim to improve the sensitivity analysis in SEMDOT to make it more mathematically rigorous. This may increase the complexity of the back end of the algorithm.

The authors hope you, the readers, have enjoyed this book. Any future improvements and suggestions will be gladly accepted by the authors.

A

Comparison with Existing Algorithms

A.1 Optimization Problem

The topology optimization algorithms can be mainly categorized into two groups: binary and relaxed formulations. The Bi-directional Evolutionary Structural Optimization (BESO) method is the typical binary formulation. The Solid Isotropic Material with Penalization (SIMP) method with the penalty coefficient of 1 (that is, $p = 1$), the so-called "variable-thickness-sheet" problem, is the fully relaxed formulation. The SIMP method with the penalty coefficient greater than 1 is the partially relaxed formulation. Although the non-penalization Smooth-Edged Material Distribution for Optimizing Topology (SEMDOT) method does not use the material penalization scheme, it should also be regarded as the partially relaxed formulation because of using the Heaviside smooth function to pure solid/void grid points that are assigned to each element. The Messerschmitt-Bölkow-Blohm (MBB) beam case, shown in Figure A.1, is taken as an example to demonstrate the difference between SEMDOT and existing algorithms such as SIMP and BESO (the soft-kill version). As BESO can only use the Optimality Criteria (OC) like optimizer, the OC optimizer is used for SIMP and SEMDOT here. It is noted that the density filter is used in SIMP. For all numerical examples, a mesh of 600×100, filter radius of 4, and target volume fraction of 0.4 are used.

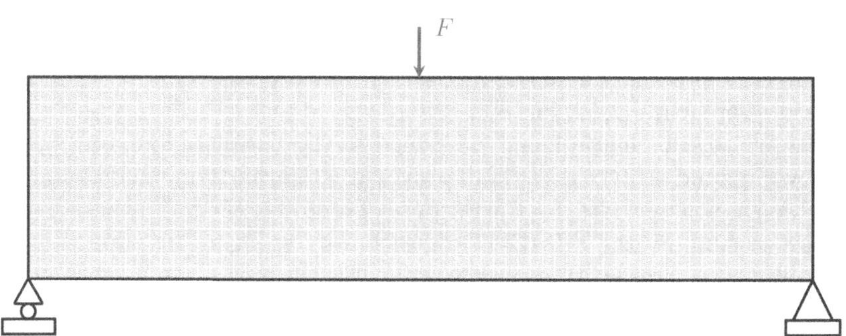

FIGURE A.1
Design domain of MBB beam case.

DOI: 10.1201/9781032634449-A

A.2 Comparison of Performance and Convergence

Structural compliance and number of iterations of BESO, SIMP with different penalty factors, and SEMDOT are summarized in Table A.1. It can be seen from Table A.1 that the objective value of SIMP with $p = 1$ (the fully relaxed formulation) is the lowest, which is 98.1309. With the increase in the penalty factor in SIMP, structural compliance increases. The objective value of BESO is close to that of SIMP with $p = 1.2$. The objective value of SEMDOT is close to that of SIMP with $p = 1.5$. BESO, the binary formulation, converges the fastest, with the total number of iterations of 55. In terms of the partially relaxed formulations, SEMDOT converges much faster than SIMP with a penalty factor greater than 1. Topological configurations of BESO, SIMP, and SEMDOT are very different. When using $p = 1$ and $p = 1.2$ in SIMP, numerous intermediate elements can be observed. Thanks to the smooth boundary, the topological configuration obtained by SEMDOT is the most practical. The post-processing methods are required for the topologies obtained by BESO and SIMP before fabrication. It is noted that the comparison of BESO with SIMP and SEMDOT is not 100% fair as the optimization solvers are slightly different.

There is an incorrect understanding that SEMDOT just uses the post-processing method for SIMP. If so, the main structures of SEMDOT and SIMP shown in Table A.1 should be

TABLE A.1
Structural compliance and total number of iterations of MBB beam case obtained by BESO, SIMP, and SEMDOT

Algorithms	Topological configuration	Compliance	Iteration
BESO		109.8329	55
SIMP with $p = 1$		98.1309	61
SIMP with $p = 1.2$	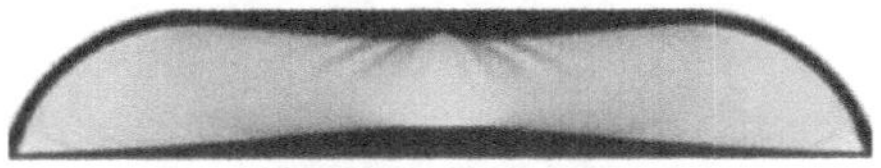	105.9310	333
SIMP with $p = 1.5$		113.4232	1338
SIMP with $p = 2$		119.3606	789
SIMP with $p = 2.5$		124.1095	1322
SIMP with $p = 3$		127.9225	997
SEMDOT		113.6358	225

similar. However, the topological configurations of these two algorithms are completely different. Therefore, SEMDOT should be considered as a new manner of pursuing the optimal material distribution.

A.3 Comparison of Convergence Process

The convergence process of one algorithm can reflect its optimization mechanism. The convergence processes of SIMP, BESO, and SEMDOT are shown in Figure A.2. Figure A.2(a) shows that the structural compliance starts from a high value and then dramatically reduces to a low value. After that, the optimization process of SIMP reaches steady state. Figure A.2(b) shows that with the increase in the volume, the structural compliance increases from a low value. When the volume constraint is met, the final compliance value of BESO is obtained. Figure A.2(c) shows that the optimization process of SEMDOT starts from a high value and then quickly reduces to the lowest point with a value of 108.0024, which is close to the value obtained by SIMP with $p = 1.2$. Afterwards, the structural compliance value increases and then gradually decreases to a constant value. In addition, the topological boundary error gradually decreases until reaching a negligible value (for example, 0.001).

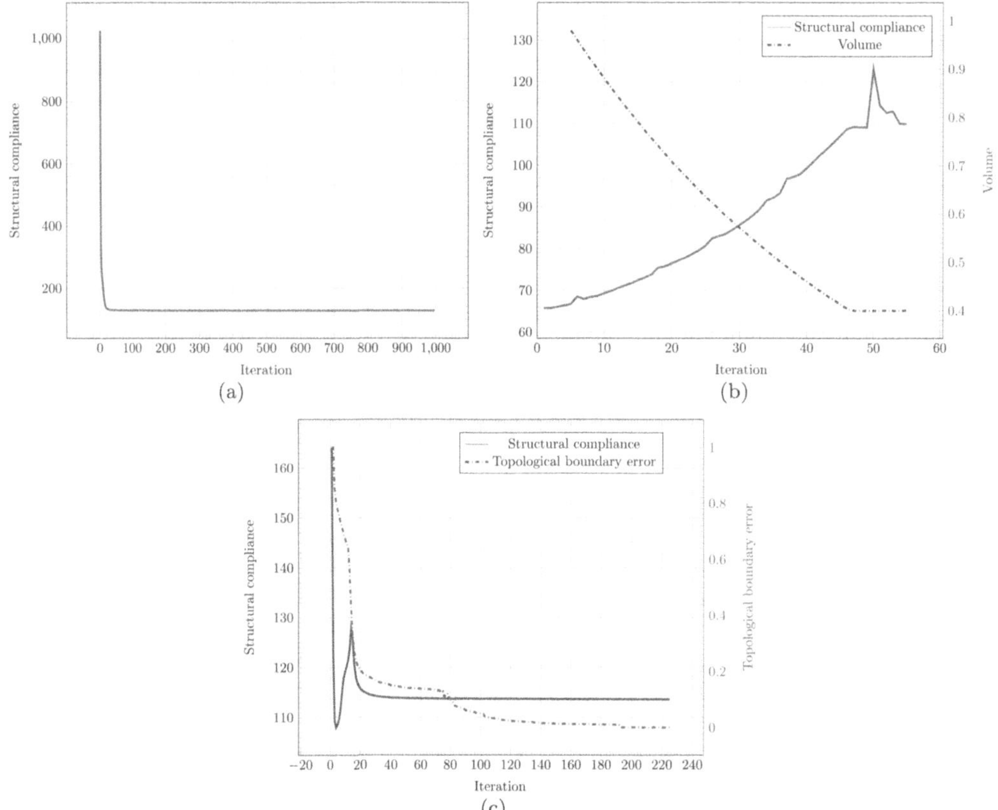

FIGURE A.2
Optimization processes of (a) SIMP, (b) BESO, and (c) SEMDOT.

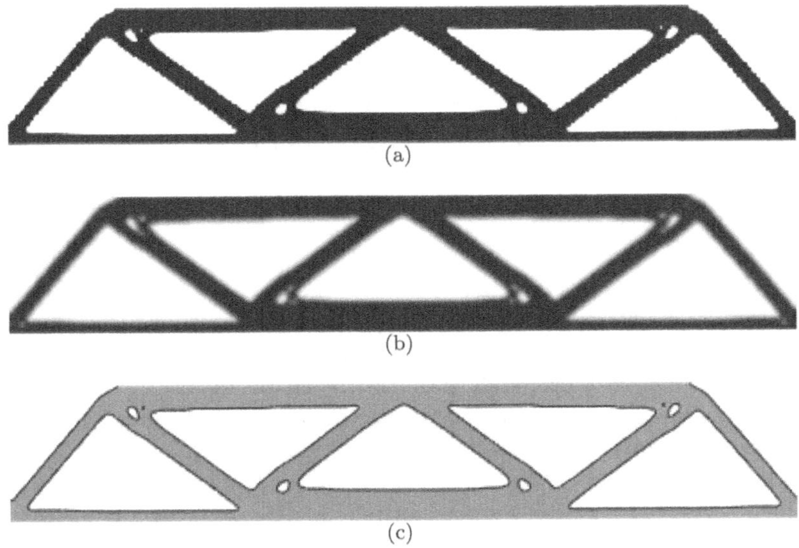

FIGURE A.3
Three levels of SEMDOT: (a) element-based, (b) grid-point-based, and (c) smooth results.

A.4 Remarks on Topological Results

SIMP and BESO only have the element-based results; however, SEMDOT has three levels of results: element-based, grid-point-based, and smooth boundary levels, as illustrated in Figure A.3. In SEMDOT, the element-based result (Figure A.3a) is used for FEA; the grid-point-based result (Figure A.3b) is the preparation for generating smooth topological boundaries; and the smooth result (Figure A.3c) is obtained based on the grid point distribution.

The SEMDOT algorithm can be regarded as a two-step approach. The first step is that the Heaviside smooth function is used to push the relaxed solution toward binary. By contrast, in SIMP, this is done by means of the power-law function. The second step is that grid points are used at the boundary elements to convert the gray elements to partial elements and obtain a smooth binary solution. Although SEMDOT is developed based on the SIMP framework, its algorithm mechanism is very different with SIMP.

B

Zhou-Rozvany Problem

A challenging benchmark example proposed by Zhou and Rozvany [219], the famous Zhou-Rozvany problem, is widely used to test the optimality of newly proposed algorithms. The Zhou-Rozvany problem was originally proposed to challenge the optimality of the Evolutionary Structural Optimization (ESO) approach. In recent years, massive efforts were made by researchers to properly solve the Zhou-Rozvany problem [58,134,166,182]. This appendix aims to provide an example of solving the Zhou-Rozvany problem with the compliance-based Smooth-Edged Material Distribution for Optimizing Topology (SEMDOT) algorithm. The design domain and boundary and loading conditions of the basic Zhou-Rozvany problem are illustrated in Figure B.1(a), where 100 four-node quadrilateral elements are taken into account, and the loading condition consists of a horizontal load of intensity 2 and a vertical load of intensity 1. In the Zhou-Rozvany problem, if one of the four elements (94-97) in the vertical rod is eliminated, the boundary conditions will be entirely changed, resulting in a non-optimal solution. When using the target volume fraction of 0.4, the resulting topology by SEMDOT is shown in Figure B.1(b), where no elements are removed from the vertical rod, showing its effectiveness in solving the Zhou-Rozvany problem. When using SIMP, the four intermediate elements are in the vertical bar, as shown in Figure B.1(c). It is hard to determine the exact material distribution in the vertical bar. However, in BESO, the four elements and the elements close to the fixed constraint are completely removed from the vertical bar, resulting in an improper design, as shown in Figure B.1(d). It seems that the Heaviside smooth function in SEMDOT is a better way to pursue the topological solution in comparison with the material penalization scheme in SIMP and BESO.

DOI: 10.1201/9781032634449-B

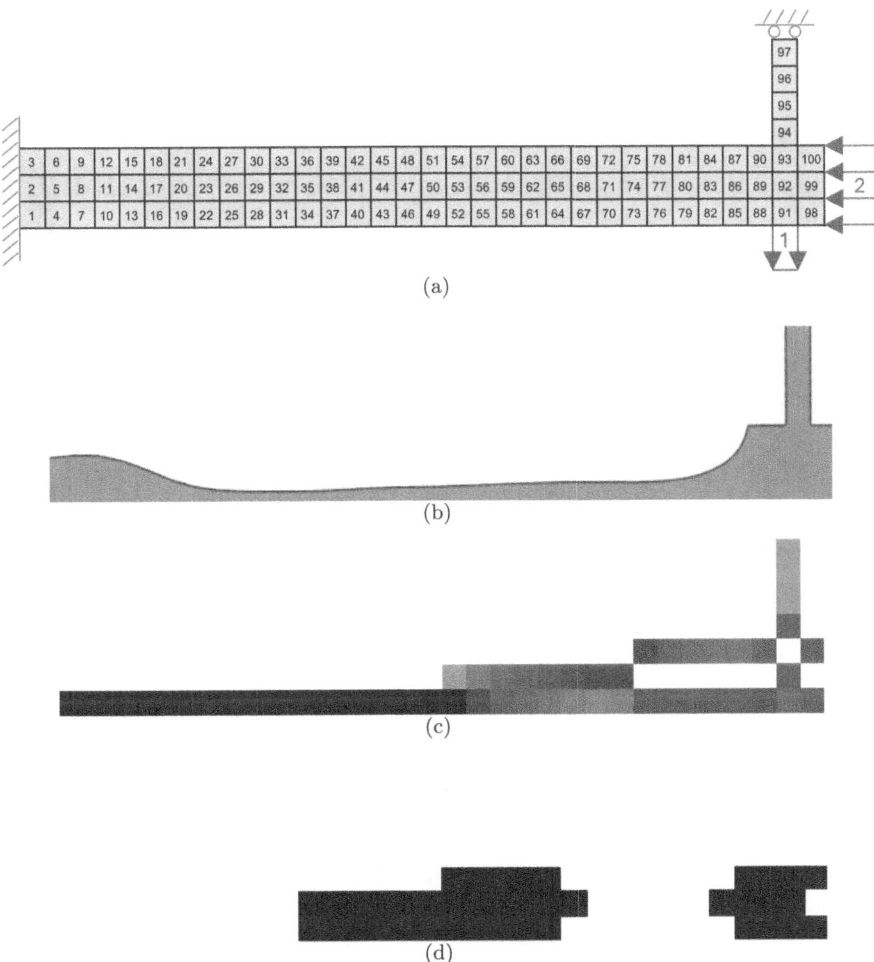

FIGURE B.1
Zhou–Rozvany problem: (a) design domain, (b) optimized structure with SEMDOT, (c) optimized structure with SIMP, and (d) optimized structure with BESO.

C

MATLAB-Based SEMDOT Software Packages

C.1 Image-Based Initialization and Post-processing (IbIPP) Package

An Image-based Initialization and Post-processing methodology (IbIPP) proposed by Osezua et al. [85] uses an input image to represent the design's features for topology optimization. The Smooth-Edged Material Distribution for Optimizing Topology (SEMDOT) algorithm is now involved in IbIPP. The design problem, input image for IbIPP, and topological configuration obtained by SEMDOT are illustrated in Figure C.1. The open-source MATLAB codes of IbIPP can be found in the link: https://github.com/ooibhadode/IbIPP.

C.2 Optiworks

Optiworks is a simple, free, and all-in-one topology optimization app, developed by Dr. Osezua Ibhadode and his graduate student Chinedu Ifediorah. Optiworks is the extension app of FreeTO (Freeform Topology Optimization). SEMDOT is a built-in topology optimization algorithm in Optiworks. Optiworks is able to initialize, mesh, and optimize arbitrary domains (see Figure C.2) and cuboid domains (see Figure C.3). The basic static analysis can be conducted to obtain the equivalent stress, equivalent strain, and displacement before and after optimization, as shown in Figure C.2(b). In addition, STereoLithography (STL) files can be generated after optimization. The software and tutorial videos can be downloaded from the link: https://sites.google.com/ualberta.ca/mdam/software?authuser=0.

DOI: 10.1201/9781032634449-C

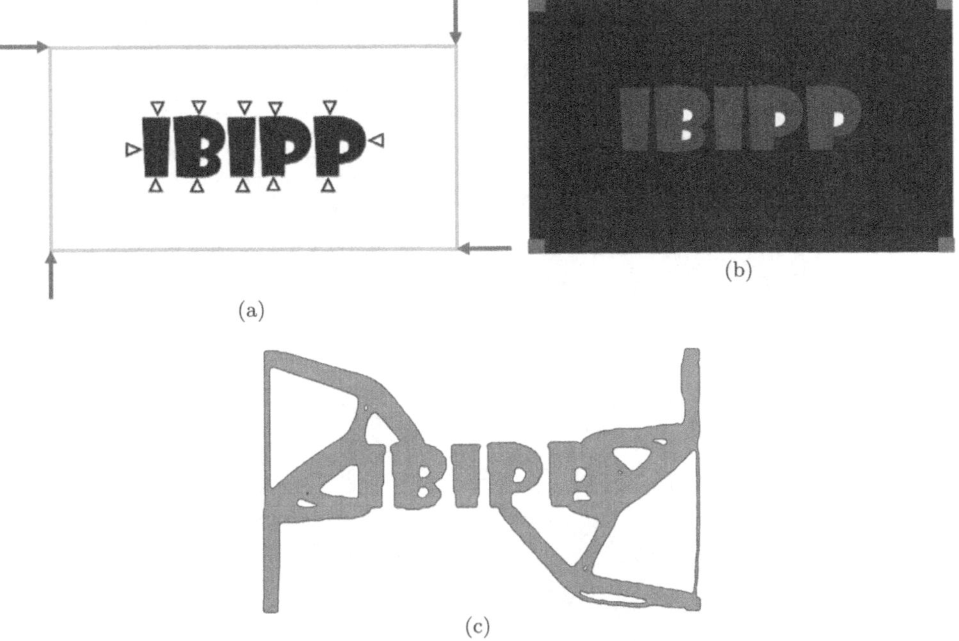

FIGURE C.1
Workflow of generating a topologically optimized structure using IbIPP: (a) design problem, (b) input image, and (c) final topology.

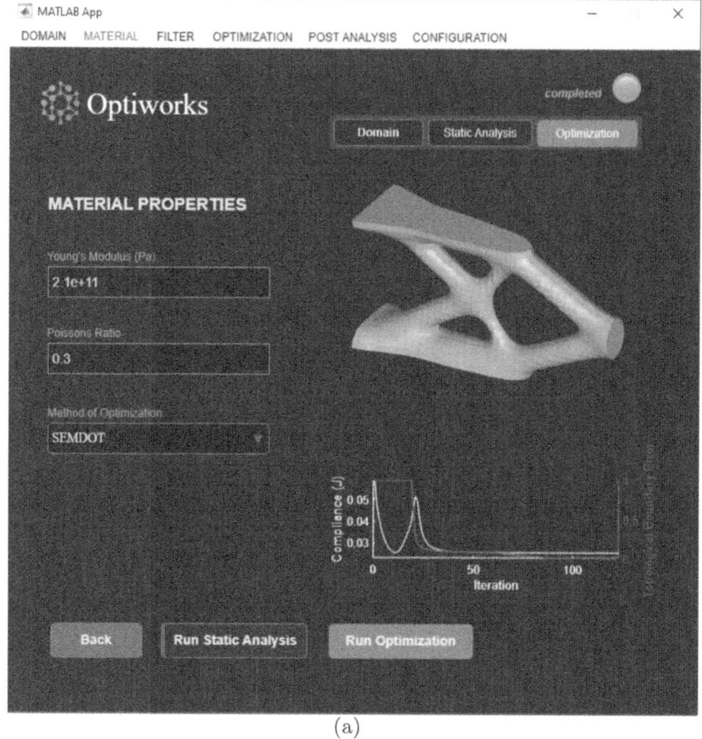

(a)

(b)

FIGURE C.2
Optiworks for cuboid domain: (a) topology optimization and (b) finite element analysis.

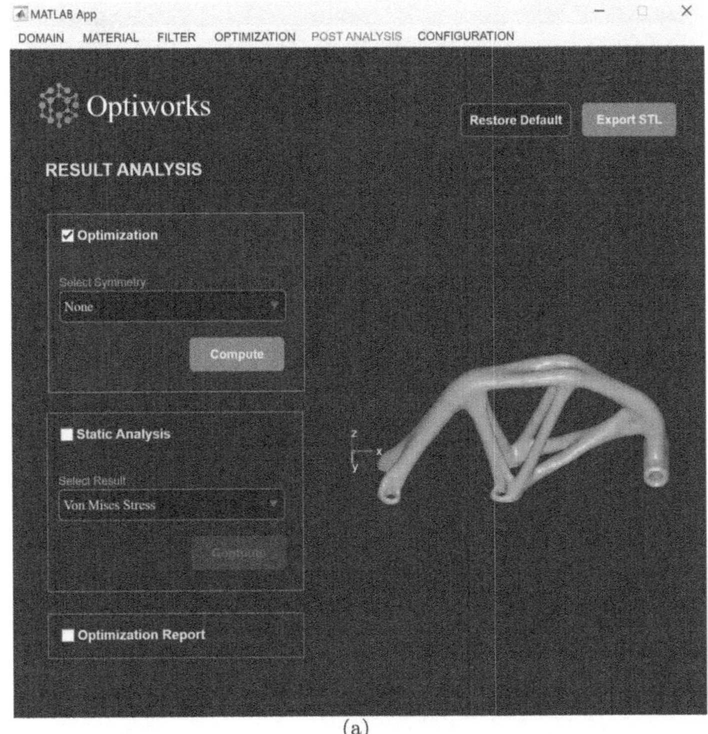

(a)

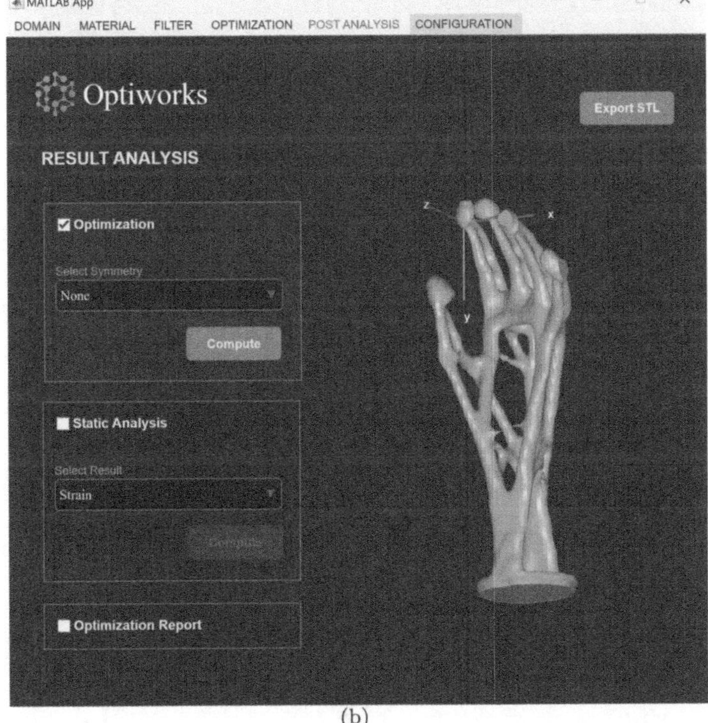

(b)

FIGURE C.3
Optiworks for arbitrary domain: (a) industrial frame and (b) hand.

Bibliography

[1] N. Aage, E. Andreassen, and B.S. Lazarov. Topology optimization using PETSc: An easy-to-use, fully parallel, open source topology optimization framework. *Structural and Multidisciplinary Optimization*, 51:565–572, 2015.

[2] D.W. Abueidda, S. Koric, and N.A. Sobh. Topology optimization of 2D structures with nonlinearities using deep learning. *Computers & Structures*, 237:106283, 2020.

[3] J. Alexandersen. A detailed introduction to density-based topology optimisation of fluid flow problems with implementation in MATLAB. *Structural and Multidisciplinary Optimization*, 66(1):12, 2023.

[4] J. Alexandersen, N. Aage, C.S. Andreasen, and O. Sigmund. Topology optimisation for natural convection problems. *International Journal for Numerical Methods in Fluids*, 76(10):699–721, 2014.

[5] J. Alexandersen and C.S. Andreasen. A review of topology optimisation for fluid-based problems. *Fluids*, 5(1):29, 2020.

[6] J. Alexandersen, O. Sigmund, and N. Aage. Large scale three-dimensional topology optimisation of heat sinks cooled by natural convection. *International Journal of Heat and Mass Transfer*, 100:876–891, 2016.

[7] G. Allaire, C. Dapogny, and P. Frey. A mesh evolution algorithm based on the level set method for geometry and topology optimization. *Structural and Multidisciplinary Optimization*, 48:711–715, 2013.

[8] G. Allaire, F. Jouve, and A.M. Toader. Structural optimization using sensitivity analysis and a level-set method. *Journal of Computational Physics*, 194(1):363–393, 2004.

[9] E. Andreassen and C.S. Andreasen. How to determine composite material properties using numerical homogenization. *Computational Materials Science*, 83:488–495, 2014.

[10] E. Andreassen, A. Clausen, M. Schevenels, B.S. Lazarov, and O. Sigmund. Efficient topology optimization in MATLAB using 88 lines of code. *Structural and Multidisciplinary Optimization*, 43:1–16, 2011.

[11] J. Asmussen, J. Alexandersen, O. Sigmund, and C.S. Andreasen. A "poor man"' approach to topology optimization of natural convection problems. *Structural and Multidisciplinary Optimization*, 59(4):1105–1124, 2019.

[12] S. Barmada, N. Fontana, A. Formisano, D. Thomopulos, and M. Tucci. A deep learning surrogate model for topology optimization. *IEEE Transactions on Magnetics*, 57(6):1–4, 2021.

[13] R. Barrett, M. Berry, T.F. Chan, J. Demmel, J. Donato, J. Dongarra, V. Eijkhout, R. Pozo, C. Romine, and H. Van der Vorst. *Templates for the solution of linear systems: Building blocks for iterative methods*. SIAM, 1994.

[14] B. Barroqueiro, A. Andrade-Campos, and R.A.F. Valente. Designing self supported SLM structures via topology optimization. *Journal of Manufacturing and Materials Processing*, 3(3):68, 2019.

[15] M. Bayat, O. Zinovieva, F. Ferrari, C. Ayas, M. Langelaar, J. Spangenberg, R. Salajeghe, K. Poulios, S. Mohanty, O. Sigmund, and J. Hattel. Holistic computational design within additive manufacturing through topology optimization combined with multiphysics multi-scale materials and process modelling. *Progress in Materials Science*, 138:101129, 2023.

[16] M.M. Behzadi and H.T. Ilieş. Real-time topology optimization in 3D via deep transfer learning. *Computer-Aided Design*, 135:103014, 2021.

[17] M.P. Bendsøe. Optimal shape design as a material distribution problem. *Structural Optimization*, 1:193–202, 1989.

[18] M.P. Bendsøe. *Optimization of structural topology, shape, and material*, vol. 414. Springer, 1995.

[19] M.P. Bendsøe and N. Kikuchi. Generating optimal topologies in structural design using a homogenization method. *Computer Methods in Applied Mechanics and Engineering*, 71(2):197–224, 1988.

[20] M.P. Bendsøe and O. Sigmund. *Topology optimization: Theory, methods, and applications*. Springer Science & Business Media, 2003.

[21] M.H. Bi, P. Tran, and Y.M. Xie. Topology optimization of 3D continuum structures under geometric self-supporting constraint. *Additive Manufacturing*, 36:101422, 2020.

[22] E. Biyikli and A.C. To. Proportional topology optimization: A new non-sensitivity method for solving stress constrained and minimum compliance problems and its implementation in MATLAB. *PLOS One*, 10(12):e0145041, 2015.

[23] M. Bruggi. On an alternative approach to stress constraints relaxation in topology optimization. *Structural and Multidisciplinary Optimization*, 36:125–141, 2008.

[24] S. Campocasso, M. Chalvin, U. Bourgon, V. Hugel, and M. Museau. Manufacturing of a Schwarz-P pattern by multi-axis WAAM. *CIRP Annals*, 72(1):377–380, 2023.

[25] V.J. Challis. A discrete level-set topology optimization code written in Matlab. *Structural and Multidisciplinary Optimization*, 41:453–464, 2010.

[26] Y. Chen, L. Ye, and X. Han. Experimental and numerical investigation of zero Poisson's ratio structures achieved by topological design and 3D printing of SCF/PA. *Composite Structures*, 293:115717, 2022.

[27] Y.F. Chen, F. Meng, G.Y. Li, and X.D. Huang. Topology optimization of photonic crystals with exotic properties resulting from dirac-like cones. *Acta Materialia*, 164:377–389, 2019.

[28] Z. Chen, K. Long, P. Wen, and S. Nouman. Fatigue-resistance topology optimization of continuum structure by penalizing the cumulative fatigue damage. *Advances in Engineering Software*, 150:102924, 2020.

[29] Z. Chen, K. Long, C. Zhang, X. Yang, F. Lu, R. Wang, B. Zhu, and X. Zhang. A fatigue-resistance topology optimization formulation for continua subject to general loads using rainflow counting. *Structural and Multidisciplinary Optimization*, 66(9):210, 2023.

[30] K.T. Cheng and N. Olhoff. An investigation concerning optimal design of solid elastic plates. *International Journal of Solids and Structures*, 17(3):305–323, 1981.

[31] L. Cheng and A. To. Part-scale build orientation optimization for minimizing residual stress and support volume for metal additive manufacturing: Theory and experimental validation. *Computer-Aided Design*, 113:1–23, 2019.

[32] C.F. Christensen, F.W. Wang, and O. Sigmund. Topology optimization of multiscale structures considering local and global buckling response. *Computer Methods in Applied Mechanics and Engineering*, 408:115969, 2023.

[33] D.C. Da, L. Xia, G.Y. Li, and X.D. Huang. Evolutionary topology optimization of continuum structures with smooth boundary representation. *Structural and Multidisciplinary Optimization*, 57:2143–2159, 2018.

[34] A.L.F. da Silva, R.A. Salas, and E.C.N. Silva. Topology optimization of fiber reinforced structures considering stress constraint and optimized penalization. *Composite Structures*, 316:117006, 2023.

[35] G.A. da Silva, N. Aage, A.T. Beck, and O. Sigmund. Local versus global stress constraint strategies in topology optimization: a comparative study. *International Journal for Numerical Methods in Engineering*, 122(21):6003–6036, 2021.

[36] G.A. da Silva, N. Aage, A.T. Beck, and O. Sigmund. Three-dimensional manufacturing tolerant topology optimization with hundreds of millions of local stress constraints. *International Journal for Numerical Methods in Engineering*, 122(2):548–578, 2021.

[37] J.D. Deaton and R.V. Grandhi. A survey of structural and multidisciplinary continuum topology optimization: Post 2000. *Structural and Multidisciplinary Optimization*, 49:1–38, 2014.

[38] A.R. Díaaz and N. Kikuchi. Solutions to shape and topology eigenvalue optimization problems using a homogenization method. *International Journal for Numerical Methods in Engineering*, 35(7):1487–1502, 1992.

[39] Z.L. Du, T.C. Cui, C. Liu, W.S. Zhang, Y.L. Guo, and X. Guo. An efficient and easy-to-extend Matlab code of the Moving Morphable Component (MMC) method for three-dimensional topology optimization. *Structural and Multidisciplinary Optimization*, 65(5):158, 2022.

[40] H.A. Eschenauer and N. Olhoff. Topology optimization of continuum structures: A review. *Applied Mechanics Reviews*, 54(4):331–390, 2001.

[41] F. Ferrari, O. Sigmund, and J.K. Guest. Topology optimization with linearized buckling criteria in 250 lines of MATLAB. *Structural and Multidisciplinary Optimization*, 63(6):3045–3066, 2021.

[42] F. Ferrari and O. Sigmund. Revisiting topology optimization with buckling constraints. *Structural and Multidisciplinary Optimization*, 59:1401–1415, 2019.

[43] F. Ferrari and O. Sigmund. A new generation 99 line Matlab code for compliance topology optimization and its extension to 3D. *Structural and Multidisciplinary Optimization*, 62:2211–2228, 2020.

[44] C. Fleury. Mathematical programming methods for constrained optimization: Dual methods. *Progress in Astronautics and Aeronautics*, 150:123–150, 1993.

[45] Y.F. Fu. Recent advances and future trends in exploring Pareto-optimal topologies and additive manufacturing oriented topology optimization. *Mathematical Biosciences and Engineering*, 17(5):4631–4656, 2020.

[46] Y.F. Fu. *Smooth Topological Design of Continuum Structures for Additive Manufacturing*. PhD thesis, Deakin University, 2021.

[47] Y.F. Fu, K. Long, and B. Rolfe. On non-penalization SEMDOT using discrete variable sensitivities. *Journal of Optimization Theory and Applications*, 198:644–677, 2023.

[48] Y.F. Fu, K. Long, A. Zolfagharian, M. Bodaghi, and B. Rolfe. Topological design of cellular structures for maximum shear modulus using homogenization SEMDOT. *Materials Today: Proceedings*, 1–5, 2023.

[49] Y.F. Fu and B. Rolfe. 4D printing roadmap: topology optimization for 4D printing. *Smart Materials and Structures*, 33:113501.

[50] Y.F. Fu and B. Rolfe. Non-penalization topology optimization for maximizing natural frequency using SEMDOT. In *Proceedings of IASS Annual Symposia*. International Association for Shell and Spatial Structures (IASS), 2505–2511, 2023.

[51] Y.F. Fu, B. Rolfe, L.N.S. Chiu, Y.N. Wang, X.D. Huang, and K. Ghabraie. Design and experimental validation of self-supporting topologies for additive manufacturing. *Virtual and Physical Prototyping*, 14(4):382–394, 2019.

[52] Y.F. Fu, B. Rolfe, L.N.S. Chiu, Y.N. Wang, X.D. Huang, and K. Ghabraie. Parametric studies and manufacturability experiments on smooth self-supporting topologies. *Virtual and Physical Prototyping*, 15(1):22–34, 2020.

[53] Y.F. Fu, B. Rolfe, L.N.S. Chiu, Y.N Wang, X.D. Huang, and K. Ghabraie. SEMDOT: Smooth-edged material distribution for optimizing topology algorithm. *Advances in Engineering Software*, 150:102921, 2020.

[54] Y.F. Fu, B. Rolfe, L.N.S. Chiu, Y.N. Wang, X.D. Huang, and K. Ghabraie. Smooth topological design of 3D continuum structures using elemental volume fractions. *Computers & Structures*, 231:106213, 2020.

[55] T. Gao, W.H. Zhang, J.H. Zhu, Y.J. Xu, and D.H. Bassir. Topology optimization of heat conduction problem involving design-dependent heat load effect. *Finite Elements in Analysis and Design*, 44(14):805–813, 2008.

[56] A. Garaigordobil, R. Ansola, and I. Fernandez de Bustos. On preventing the dripping effect of overhang constraints in topology optimization for additive manufacturing. *Structural and Multidisciplinary Optimization*, 64(6):4065–4078, 2021.

[57] A.T. Gaynor and J.K. Guest. Topology optimization considering overhang constraints: Eliminating sacrificial support material in additive manufacturing through design. *Structural and Multidisciplinary Optimization*, 54(5):1157–1172, 2016.

[58] K. Ghabraie. The ESO method revisited. *Structural and Multidisciplinary Optimization*, 51:1211–1222, 2015.

[59] L.V. Gibiansky and O. Sigmund. Multiphase composites with extremal bulk modulus. *Journal of the Mechanics and Physics of Solids*, 48(3):461–498, 2000.

[60] I. Gibson, D.W. Rosen, B. Stucker, and M. Khorasani. *Additive manufacturing technologies*, vol. 17. Springer, 2021.

[61] P.E. Gill, W. Murray, and M.A. Saunders. SNOPT: An SQP algorithm for large-scale constrained optimization. *SIAM Review*, 47(1):99–131, 2005.

[62] O. Giraldo-Londoño and G.H. Paulino. PolyStress: A MATLAB implementation for local stress-constrained topology optimization using the augmented Lagrangian method. *Structural and Multidisciplinary Optimization*, 63:2065–2097, 2021.

[63] J.P. Groen, C.R. Thomsen, and O. Sigmund. Multi-scale topology optimization for stiffness and de-homogenization using implicit geometry modeling. *Structural and Multidisciplinary Optimization*, 63(6):2919–2934, 2021.

[64] A.A. Groenwold and L.F.P. Etman. A simple heuristic for gray-scale suppression in optimality criterion-based topology optimization. *Structural and Multidisciplinary Optimization*, 39:217–225, 2009.

[65] A.A. Groenwold and L.F.P. Etman. A quadratic approximation for structural topology optimization. *International Journal for Numerical Methods in Engineering*, 82(4):505–524, 2010.

[66] J.K. Guest, J.H. Prévost, and T. Belytschko. Achieving minimum length scale in topology optimization using nodal design variables and projection functions. *International Journal for Numerical Methods in Engineering*, 61(2):238–254, 2004.

[67] X. Guo, W.S. Zhang, and W.L. Zhong. Doing topology optimization explicitly and geometrically—A new moving morphable components based framework. *Journal of Applied Mechanics*, 81(8):081009, 2014.

[68] R. Haber and M.P. Bendsøe. Problem formulation, solution procedures and geometric modeling-key issues in variable-topology optimization. In *7th AIAA/USAF/NASA/ISSMO Symposium on Multidisciplinary Analysis and Optimization*, 4948, 1998.

[69] B. Hassani and E. Hinton. A review of homogenization and topology opimization II—Analytical and numerical solution of homogenization equations. *Computers & Structures*, 69(6):719–738, 1998.

[70] B. Hassani and E. Hinton. A review of homogenization and topology optimization I—Homogenization theory for media with periodic structure. *Computers & Structures*, 69(6):707–717, 1998.

[71] B. Hassani and E. Hinton. *Homogenization and structural topology optimization: Theory, practice and software*. Springer Science & Business Media, 2012.

[72] E. Holmberg, B. Torstenfelt, and A. Klarbring. Stress constrained topology optimization. *Structural and Multidisciplinary Optimization*, 48:33–47, 2013.

[73] X.D. Huang. Smooth topological design of structures using the floating projection. *Engineering Structures*, 208:110330, 2020.

[74] X.D. Huang. On smooth or 0/1 designs of the fixed-mesh element-based topology optimization. *Advances in Engineering Software*, 151:102942, 2021.

[75] X.D. Huang and W.B. Li. Three-field floating projection topology optimization of continuum structures. *Computer Methods in Applied Mechanics and Engineering*, 399:115444, 2022.

[76] X.D. Huang, W.B. Li, K. Nabaki, and X.L. Yan. Reformulation for stress topology optimization of continuum structures by floating projection. *Computer Methods in Applied Mechanics and Engineering*, 423:116870, 2024.

[77] X.D. Huang, Y. Li, S.W. Zhou, and Y.M. Xie. Topology optimization of compliant mechanisms with desired structural stiffness. *Engineering Structures*, 79:13–21, 2014.

[78] X.D. Huang, A. Radman, and Y.M. Xie. Topological design of microstructures of cellular materials for maximum bulk or shear modulus. *Computational Materials Science*, 50(6):1861–1870, 2011.

[79] X.D. Huang and Y.M. Xie. Convergent and mesh-independent solutions for the bi-directional evolutionary structural optimization method. *Finite Elements in Analysis and Design*, 43(14):1039–1049, 2007.

[80] X.D. Huang and Y.M. Xie. Bi-directional evolutionary topology optimization of continuum structures with one or multiple materials. *Computational Mechanics*, 43:393–401, 2009.

[81] X.D. Huang and Y.M. Xie. A further review of ESO type methods for topology optimization. *Structural and Multidisciplinary Optimization*, 41:671–683, 2010.

[82] X.D. Huang, Z.H. Zuo, and Y.M. Xie. Evolutionary topological optimization of vibrating continuum structures for natural frequencies. *Computers & Structures*, 88(5-6):357–364, 2010.

[83] O. Ibhadode, Y.F. Fu, and A. Qureshi. FreeTO: Freeform 3D topology optimization using a structured mesh with smooth boundaries in MATLAB. *Advances in Engineering Software*, 198:103790, 2024.

[84] O. Ibhadode, Z. Zhang, J. Sixt, K.M. Nsiempba, J. Orakwe, A. Martinez-Marchese, O. Ero, S.I. Shahabad, A. Bonakdar, and E. Toyserkani. Topology optimization for metal additive manufacturing: Current trends, challenges, and future outlook. *Virtual and Physical Prototyping*, 18(1):e2181192, 2023.

[85] O. Ibhadode, Z.D. Zhang, A. Bonakdar, and E. Toyserkani. IbIPP for topology optimization—An image-based initialization and post-processing code written in MATLAB. *SoftwareX*, 14:100701, 2021.

[86] J.C. Jiang and Y.F. Fu. A short survey of sustainable material extrusion additive manufacturing. *Australian Journal of Mechanical Engineering*, 21(1):123–132, 2023.

[87] J.C. Jiang, X. Xu, and J. Stringer. Support structures for additive manufacturing: A review. *Journal of Manufacturing and Materials Processing*, 2(4):64, 2018.

[88] M. Kato, T. Kii, K. Yaji, and K. Fujita. Tackling an exact maximum stress minimization problem with gradient-free topology optimization incorporating a deep generative model. In *International Design Engineering Technical Conferences and Computers and Information in Engineering Conference*, vol. 87318. American Society of Mechanical Engineers, V03BT03A008, 2023.

[89] G.A. Katsifis, D.R. McKenzie, and N. Suchowerska. Applying the Hashin–Shtrikman bounds to predict stiffness of multicomponent 3D printed structures: Towards regenerative orthopaedic medicine. *Journal of Composite Materials*, 54(16):2173–2183, 2020.

[90] T. Kii, K. Yaji, K. Fujita, Z. Sha, and C.C. Seepersad. Latent crossover for data-driven multifidelity topology design. *Journal of Mechanical Design*, 146(5):051713, 2024.

[91] C.Y. Kiyono, S.L. Vatanabe, E.C.N. Silva, and J.N. Reddy. A new multi-p-norm formulation approach for stress-based topology optimization design. *Composite Structures*, 156:10–19, 2016.

[92] C. Kocer, D.R. McKenzie, and M.M. Bilek. Elastic properties of a material composed of alternating layers of negative and positive Poisson's ratio. *Materials Science and Engineering: A*, 505(1–2):111–115, 2009.

[93] M. Langelaar. Topology optimization of 3D self-supporting structures for additive manufacturing. *Additive Manufacturing*, 12:60–70, 2016.

[94] M. Langelaar. An additive manufacturing filter for topology optimization of print-ready designs. *Structural and Multidisciplinary Optimization*, 55:871–883, 2017.

[95] M. Langelaar. Combined optimization of part topology, support structure layout and build orientation for additive manufacturing. *Structural and Multidisciplinary Optimization*, 57(5):1985–2004, 2018.

[96] M. Langelaar. Integrated component-support topology optimization for additive manufacturing with post-machining. *Rapid Prototyping Journal*, 25(2):255–265, 2019.

[97] B.S. Lazarov and O. Sigmund. Filters in topology optimization based on Helmholtz-type differential equations. *International Journal for Numerical Methods in Engineering*, 86(6):765–781, 2011.

[98] C. Le, J. Norato, T. Bruns, C. Ha, and D. Tortorelli. Stress-based topology optimization for continua. *Structural and Multidisciplinary Optimization*, 41:605–620, 2010.

[99] D.Z. Li and Il Y. Kim. Multi-material topology optimization for practical lightweight design. *Structural and Multidisciplinary Optimization*, 58:1081–1094, 2018.

[100] H. Li, T. Kondoh, P. Jolivet, K. Furuta, T. Yamada, B.L. Zhu, K. Izui, and S. Nishiwaki. Three-dimensional topology optimization of a fluid–structure system using body-fitted mesh adaption based on the level-set method. *Applied Mathematical Modelling*, 101:276–308, 2022.

[101] H. Li, T. Kondoh, P. Jolivet, K. Furuta, T. Yamada, B.L. Zhu, H. Zhang, K. Izui, and S. Nishiwaki. Optimum design and thermal modeling for 2D and 3D natural convection problems incorporating level set-based topology optimization with body-fitted mesh. *International Journal for Numerical Methods in Engineering*, 123(9):1954–1990, 2022.

[102] H. Li, T. Kondoh, P. Jolivet, N. Nakayama, K. Furuta, H. Zhang, B.L. Zhu, K. Izui, and S. Nishiwaki. Topology optimization for lift–drag problems incorporated with distributed unstructured mesh adaptation. *Structural and Multidisciplinary Optimization*, 65(8):222, 2022.

[103] H. Li, T. Yamada, P. Jolivet, K. Furuta, T. Kondoh, K. Izui, and S. Nishiwaki. Full-scale 3D structural topology optimization using adaptive mesh refinement based on the level-set method. *Finite Elements in Analysis and Design*, 194:103561, 2021.

[104] H. Li, M.H. Yu, P. Jolivet, J. Alexandersen, T. Kondoh, T.N. Hu, K. Furuta, K. Izui, and S. Nishiwaki. Reaction–diffusion equation driven topology optimization of high-resolution and feature-rich structures using unstructured meshes. *Advances in Engineering Software*, 180:103457, 2023.

[105] L. Li and K. Khandelwal. Two-point gradient-based MMA (TGMMA) algorithm for topology optimization. *Computers & Structures*, 131:34–45, 2014.

[106] Q.H. Li, O. Sigmund, J.S. Jensen, and N. Aage. Reduced-order methods for dynamic problems in topology optimization: A comparative study. *Computer Methods in Applied Mechanics and Engineering*, 387:114149, 2021.

[107] W.B. Li and X.D. Huang. Lightweight optimization design of structures with multiple cellular materials. *International Journal of Applied Mechanics*, 14(09):2250059, 2022.

[108] X.Q. Li, Q.H. Zhao, K. Long, and H.X. Zhang. Multi-material topology optimization of transient heat conduction structure with functional gradient constraint. *International Communications in Heat and Mass Transfer*, 131:105845, 2022.

[109] Z. Li, T.U. Lee, Y. Yao, and Y.M. Xie. Smoothing topology optimization results using pre-built lookup tables. *Advances in Engineering Software*, 173:103204, 2022.

[110] Y. Liang and G.D. Cheng. Topology optimization via sequential integer programming and Canonical relaxation algorithm. *Computer Methods in Applied Mechanics and Engineering*, 348:64–96, 2019.

[111] Y. Liang and G.D. Cheng. Further elaborations on topology optimization via sequential integer programming and Canonical relaxation algorithm and 128-line MATLAB code. *Structural and Multidisciplinary Optimization*, 61:411–431, 2020.

[112] Y. Liang, K. Sun, and G.D. Cheng. Discrete variable topology optimization for compliant mechanism design via Sequential Approximate Integer Programming with Trust Region (SAIP-TR). *Structural and Multidisciplinary Optimization*, 62:2851–2879, 2020.

[113] Y. Liang, X.Y. Yan, and G.D. Cheng. Explicit control of 2D and 3D structural complexity by discrete variable topology optimization method. *Computer Methods in Applied Mechanics and Engineering*, 389:114302, 2022.

[114] B. Liu, X. Feng, and S.M. Zhang. The effective Young's modulus of composites beyond the Voigt estimation due to the Poisson effect. *Composites Science and Technology*, 69(13):2198–2204, 2009.

[115] B.S. Liu, D. Guo, C. Jiang, G.Y. Li, and X.D. Huang. Stress optimization of smooth continuum structures based on the distortion strain energy density. *Computer Methods in Applied Mechanics and Engineering*, 343:276–296, 2019.

[116] J.K. Liu. Piecewise length scale control for topology optimization with an irregular design domain. *Computer Methods in Applied Mechanics and Engineering*, 351:744–765, 2019.

[117] J.K. Liu, A.T. Gaynor, S.K. Chen, Z. Kang, K. Suresh, A. Takezawa, L. Li, J. Kato, J.Y. Tang, C.C.L. Wang, L. Cheng, X. Liang, and A.C. To. Current and future trends in topology optimization for additive manufacturing. *Structural and Multidisciplinary Optimization*, 57(6):2457–2483, 2018.

[118] K. Liu and A. Tovar. An efficient 3D topology optimization code written in MATLAB. *Structural and Multidisciplinary Optimization*, 50:1175–1196, 2014.

[119] M. Liu, Y. Wang, Q. Wei, X. Ma, K. Zhang, X. Li, C. Bao, and B. Du. Topology optimization for reducing stress shielding in cancellous bone scaffold. *Computers & Structures*, 288:107132, 2023.

[120] S. Liu, Q. Li, W. Chen, L. Tong, and G. Cheng. An identification method for enclosed voids restriction in manufacturability design for additive manufacturing structures. *Frontiers of Mechanical Engineering*, 10:126–137, 2015.

[121] Y. Liu, L. Chen, Z. Li, and J. Du. On the global optimum for heat conduction. *International Journal of Heat and Mass Transfer*, 198:123381, 2022.

[122] J. Lógó and H. Ismail. Milestones in the 150-year history of topology optimization: A review. *Computer Assisted Methods in Engineering and Science*, 27(2–3):97–132, 2020.

[123] K. Long, A. Saeed, J. Zhang, Y. Diaeldin, F. Lu, T. Tao, Y. Li, P. Sun, and J. Yan. An overview of sequential approximation in topology optimization of continuum structure. *Computer Modeling in Engineering and Sciences*, 139(1):43–67, 2024.

[124] K. Long, X. Wang, and X.G. Gu. Multi-material topology optimization for the transient heat conduction problem using a sequential quadratic programming algorithm. *Engineering Optimization*, 50(12):2091–2107, 2018.

[125] K. Long, X. Wang, and H.L. Liu. Stress-constrained topology optimization of continuum structures subjected to harmonic force excitation using sequential quadratic programming. *Structural and Multidisciplinary Optimization*, 59:1747–1759, 2019.

[126] K. Long, X. Wang, P.W. Sun, and X. Liu. *Topology optimization methods and applications for continuum structures.* China Water and Power Press, 2022.

[127] K. Long, X.Y. Yang, N. Saeed, R.H. Tian, P. Wen, and X. Wang. Topology optimization of transient problem with maximum dynamic response constraint using soar scheme. *Frontiers of Mechanical Engineering*, 16(3):593–606, 2021.

[128] Y.J. Luo, M.Y. Wang, and Z. Kang. An enhanced aggregation method for topology optimization with local stress constraints. *Computer Methods in Applied Mechanics and Engineering*, 254:31–41, 2013.

[129] J.C. Maxwell. On reciprocal figures, frames, and diagrams of forces. *Earth and Environmental Science Transactions of the Royal Society of Edinburgh*, 26(1):1–40, 1870.

[130] A.G.M. Michell. The limits of economy of material in frame-structures. *The London, Edinburgh, and Dublin Philosophical Magazine and Journal of Science*, 8(47):589–597, 1904.

[131] M. Mohseni and S. Khodaygan. Design for additive manufacturing of topology-optimized structures based on deep learning and transfer learning. *Rapid Prototyping Journal*, 30:1411–1433, 2024.

[132] A. Moussa, S. Rahman, M. Xu, M. Tanzer, and D. Pasini. Topology optimization of 3D-printed structurally porous cage for acetabular reinforcement in total hip arthroplasty. *Journal of the Mechanical Behavior of Biomedical Materials*, 105:103705, 2020.

[133] S. Mukherjee, D. Lu, B. Raghavan, P. Breitkopf, S. Dutta, M. Xiao, and W. Zhang. Accelerating large-scale topology optimization: State-of-the-art and challenges. *Archives of Computational Methods in Engineering*, 28:1–23, 2021.

[134] D.J. Munk, G.A. Vio, and G.P. Steven. A bi-directional evolutionary structural optimisation algorithm with an added connectivity constraint. *Finite Elements in Analysis and Design*, 131:25–42, 2017.

[135] B. Niu, X.M. He, Y. Shan, and R. Yang. On objective functions of minimizing the vibration response of continuum structures subjected to external harmonic excitation. *Structural and Multidisciplinary Optimization*, 57:2291–2307, 2018.

[136] J.A. Norato, H.A. Smith, J.D. Deaton, and R.M. Kolonay. A maximum-rectifier-function approach to stress-constrained topology optimization. *Structural and Multidisciplinary Optimization*, 65(10):286, 2022.

[137] N. Olhoff and J.B. Du. Generalized incremental frequency method for topological designof continuum structures for minimum dynamic compliance subject to forced vibration at a prescribed low or high value of the excitation frequency. *Structural and Multidisciplinary Optimization*, 54:1113–1141, 2016.

[138] J. Olsen and II Y. Kim. Design for additive manufacturing: 3D simultaneous topology and build orientation optimization. *Structural and Multidisciplinary Optimization*, 62(4):1989–2009, 2020.

[139] M. Osanov and J.K. Guest. Topology optimization for architected materials design. *Annual Review of Materials Research*, 46:211–233, 2016.

[140] S. Osher and J.A. Sethian. Fronts propagating with curvature-dependent speed: Algorithms based on Hamilton-Jacobi formulations. *Journal of Computational Physics*, 79(1):12–49, 1988.

[141] M. Otomori, T. Yamada, K. Izui, and S. Nishiwaki. MATLAB code for a level set-based topology optimization method using a reaction diffusion equation. *Structural and Multidisciplinary Optimization*, 51:1159–1172, 2015.

[142] A. Panesar, M. Abdi, D. Hickman, and I. Ashcroft. Strategies for functionally graded lattice structures derived using topology optimisation for additive manufacturing. *Additive Manufacturing*, 19:81–94, 2018.

[143] J. Park and A. Sutradhar. A multi-resolution method for 3D multi-material topology optimization. *Computer Methods in Applied Mechanics and Engineering*, 285:571–586, 2015.

[144] J. Petersson. A finite element analysis of optimal variable thickness sheets. *SIAM Journal on Numerical Analysis*, 36(6):1759–1778, 1999.

[145] R. Picelli, S. Townsend, C. Brampton, J. Norato, and H.A. Kim. Stress-based shape and topology optimization with the level set method. *Computer Methods in Applied Mechanics and Engineering*, 329:1–23, 2018.

[146] X.P. Qian. Undercut and overhang angle control in topology optimization: A density gradient based integral approach. *International Journal for Numerical Methods in Engineering*, 111(3):247–272, 2017.

[147] R. Ranjan, Z. Chen, C. Ayas, M. Langelaar, and F. Van Keulen. Overheating control in additive manufacturing using a 3D topology optimization method and experimental validation. *Additive Manufacturing*, 61:103339, 2023.

[148] R. Ranjan, R. Samant, and S. Anand. Integration of design for manufacturing methods with topology optimization in additive manufacturing. *Journal of Manufacturing Science and Engineering*, 139(6):061007, 2017.

[149] T. Ribeiro, L. Bernardo, R. Carrazedo, and D. De Domenico. Topology optimisation of steel connections under compression assisted by physical and geometrical nonlinear finite element analysis and its application to an industrial case study. *Structural and Multidisciplinary Optimization*, 67(6):88, 2024.

[150] T. Ribeiro, L. Bernardo, M.C.S. Nepomuceno, N. Maugeri, P. Longo, and D. De Domenico. Experimental results for topologically optimised steel joints under tension. *Results in Engineering*, 21:101732, 2024.

[151] T. Ribeiro, Y.F. Fu, L. Bernardo, and B. Rolfe. Topology optimisation of structural steel with non-penalisation SEMDOT: Optimisation, physical nonlinear analysis, and benchmarking. *Applied Sciences*, 13(20):11370, 2023.

[152] G.I.N. Rozvany. A critical review of established methods of structural topology optimization. *Structural and Multidisciplinary Optimization*, 37:217–237, 2009.

[153] G.I.N. Rozvany, M. Zhou, and T. Birker. Generalized shape optimization without homogenization. *Structural Optimization*, 4:250–252, 1992.

[154] S. Shojaee and M. Mohammadian. Structural topology optimization using an enhanced level set method. *Scientia Iranica*, 19(5):1157–1167, 2012.

[155] O. Sigmund. Materials with prescribed constitutive parameters: An inverse homogenization problem. *International Journal of Solids and Structures*, 31(17):2313–2329, 1994.

[156] O. Sigmund. Tailoring materials with prescribed elastic properties. *Mechanics of Materials*, 20(4):351–368, 1995.

[157] O. Sigmund. On the design of compliant mechanisms using topology optimization. *Journal of Structural Mechanics*, 25(4):493–524, 1997.

[158] O. Sigmund. A 99 line topology optimization code written in MATLAB. *Structural and Multidisciplinary Optimization*, 21:120–127, 2001.

[159] O. Sigmund. Morphology-based black and white filters for topology optimization. *Structural and Multidisciplinary Optimization*, 33:401–424, 2007.

[160] O. Sigmund. On the usefulness of non-gradient approaches in topology optimization. *Structural and Multidisciplinary Optimization*, 43:589–596, 2011.

[161] O. Sigmund. On benchmarking and good scientific practise in topology optimization. *Structural and Multidisciplinary Optimization*, 65(11):315, 2022.

[162] O. Sigmund, N. Aage, and E. Andreassen. On the (non-) optimality of Michell structures. *Structural and Multidisciplinary Optimization*, 54:361–373, 2016.

[163] O. Sigmund and K. Maute. Topology optimization approaches: A comparative review. *Structural and Multidisciplinary Optimization*, 48(6):1031–1055, 2013.

[164] O. Sigmund and J. Petersson. Numerical instabilities in topology optimization: A survey on procedures dealing with checkerboards, mesh-dependencies and local minima. *Structural Optimization*, 16:68–75, 1998.

[165] T. Smit, S. Koppen, S.J. Ferguson, and B. Helgason. Conceptual design of compliant bone scaffolds by full-scale topology optimization. *Journal of the Mechanical Behavior of Biomedical Materials*, 143:105886, 2023.

[166] M. Stolpe and M.P. Bendsøe. Global optima for the Zhou–Rozvany problem. *Structural and Multidisciplinary Optimization*, 43:151–164, 2011.

[167] M. Stolpe and K. Svanberg. On the trajectories of penalization methods for topology optimization. *Structural and Multidisciplinary Optimization*, 21:128–139, 2001.

[168] K. Sun, Y. Liang, and G.D. Cheng. Sensitivity analysis of discrete variable topology optimization. *Structural and Multidisciplinary Optimization*, 65(8):216, 2022.

[169] K. Svanberg. The method of moving asymptotes—A new method for structural optimization. *International Journal for Numerical Methods in Engineering*, 24(2):359–373, 1987.

[170] K. Svanberg. A class of globally convergent optimization methods based on conservative convex separable approximations. *SIAM Journal on Optimization*, 12(2):555–573, 2002.

[171] K. Svanberg. MMA and GCMMA—Two methods for nonlinear optimization. *Technical Report, Optimization and Systems Theory*, 1: 1–15, 2007.

[172] J.W. Tian, M.Q. Li, Z.H. Han, Y. Chen, X.F.D. Gu, Q.J. Ge, and S.K. Chen. Conformal topology optimization of multi-material ferromagnetic soft active structures using an extended level set method. *Computer Methods in Applied Mechanics and Engineering*, 389:114394, 2022.

[173] S. Torquato, L.V. Gibiansky, M.J. Silva, and L.J. Gibson. Effective mechanical and transport properties of cellular solids. *International Journal of Mechanical Sciences*, 40(1):71–82, 1998.

[174] J. van der Zwet, A. Delissen, and M. Langelaar. Prevention of enclosed voids in topology optimization using a cumulative sum flood fill algorithm. *Advances in Engineering Software*, 186:103530, 2023.

[175] N.P. Van Dijk, K. Maute, M. Langelaar, and F. Van Keulen. Level-set methods for structural topology optimization: A review. *Structural and Multidisciplinary Optimization*, 48:437–472, 2013.

[176] P. Vogiatzis, S. Chen, and C. Zhou. An open source framework for integrated additive manufacturing and level-set-based topology optimization. *Journal of Computing and Information Science in Engineering*, 17(4):041012, 2017.

[177] A. Wächter and L.T. Biegler. On the implementation of an interior-point filter line-search algorithm for large-scale nonlinear programming. *Mathematical Programming*, 106:25–57, 2006.

[178] C. Wang, Z. Zhao, M. Zhou, O. Sigmund, and X.J.S. Zhang. A comprehensive review of educational articles on structural and multidisciplinary optimization. *Structural and Multidisciplinary Optimization*, 64:1–54, 2021.

[179] C.F. Wang and X.P. Qian. Simultaneous optimization of build orientation and topology for additive manufacturing. *Additive Manufacturing*, 34:101246, 2020.

[180] F.W. Wang, B.S. Lazarov, and O. Sigmund. On projection methods, convergence and robust formulations in topology optimization. *Structural and Multidisciplinary Optimization*, 43:767–784, 2011.

[181] H. Wang, W. Cheng, R. Du, S. Wang, and Y. Wang. Improved proportional topology optimization algorithm for solving minimum compliance problem. *Structural and Multidisciplinary Optimization*, 62:475–493, 2020.

[182] L.J. Wang, W.H. Yue, and M.Q. Zhu. An improved evolutionary structure optimization method for smooth topology design of structures. *International Journal of Computational Methods*, 20(04):2250061, 2023.

[183] M.Y. Wang, X.M. Wang, and D.M. Guo. A level set method for structural topology optimization. *Computer Methods in Applied Mechanics and Engineering*, 192(1–2):227–246, 2003.

[184] S. Wang, E. de Sturler, and G.H. Paulino. Large-scale topology optimization using preconditioned Krylov subspace methods with recycling. *International Journal for Numerical Methods in Engineering*, 69(12):2441–2468, 2007.

[185] X. Wang, H.L. Liu, Z. Kang, K. Long, and Z. Meng. Topology optimization for minimum stress design with embedded movable holes. *Computers & Structures*, 244:106455, 2021.

[186] Y.G. Wang and Z. Kang. MATLAB implementations of velocity field level set method for topology optimization: An 80-line code for 2D and a 100-line code for 3D problems. *Structural and Multidisciplinary Optimization*, 64:4325–4342, 2021.

[187] Z. Wang, A. Srinivasa, J.N. Reddy, and A. Dubrowski. Topology optimization of lightweight structures with application to bone scaffolds and 3D printed shoes for diabetics. *Journal of Applied Mechanics*, 89(4):041009, 2022.

[188] P. Wei, Z.Y. Li, X.P. Li, and M.Y. Wang. An 88-line MATLAB code for the parameterized level set method based topology optimization using radial basis functions. *Structural and Multidisciplinary Optimization*, 58:831–849, 2018.

[189] R.V. Woldseth, N. Aage, J.A. Bærentzen, and O. Sigmund. On the use of artificial neural networks in topology optimisation. *Structural and Multidisciplinary Optimization*, 65(10):294, 2022.

[190] J. Wu, O. Sigmund, and J.P. Groen. Topology optimization of multi-scale structures: A review. *Structural and Multidisciplinary Optimization*, 63:1455–1480, 2021.

[191] S.H. Wu, Y.C. Zhang, and S.T. Liu. Topology optimization for minimizing the maximum temperature of transient heat conduction structure. *Structural and Multidisciplinary Optimization*, 60:69–82, 2019.

[192] S.H. Wu, Y.C. Zhang, and S.T. Liu. Transient thermal dissipation efficiency based method for topology optimization of transient heat conduction structures. *International Journal of Heat and Mass Transfer*, 170:121004, 2021.

[193] X.M. Wu, Y. Zhang, L. Gao, and J. Gao. On the indispensability of isogeometric analysis in topology optimization for smooth or binary designs. *Symmetry*, 14(5):845, 2022.

[194] Z.J. Wu and R.B. Xiao. A topology optimization approach to structure design with self-supporting constraints in additive manufacturing. *Journal of Computational Design and Engineering*, 9(2):364–379, 2022.

[195] L. Xia and P. Breitkopf. Design of materials using topology optimization and energy-based homogenization approach in MATLAB. *Structural and Multidisciplinary Optimization*, 52(6):1229–1241, 2015.

[196] L. Xia, Q. Xia, X.D. Huang, and Y.M. Xie. Bi-directional evolutionary structural optimization on advanced structures and materials: A comprehensive review. *Archives of Computational Methods in Engineering*, 25:437–478, 2018.

[197] Y.M. Xie and G.P. Steven. A simple evolutionary procedure for structural optimization. *Computers and Structures*, 49(5):885–896, 1993.

[198] Y.L. Xiong, S. Yao, Z.L. Zhao, and Y.M. Xie. A new approach to eliminating enclosed voids in topology optimization for additive manufacturing. *Additive Manufacturing*, 32:101006, 2020.

[199] B. Xu, Y.S. Han, L. Zhao, and Y.M. Xie. Topology optimization of continuum structures for natural frequencies considering casting constraints. *Engineering Optimization*, 51(6):941–960, 2019.

[200] S.Z. Xu, J.K. Liu, J.Q. Huang, B. Zou, and Y.S. Ma. Multi-scale topology optimization with shell and interface layers for additive manufacturing. *Additive Manufacturing*, 37:101698, 2021.

[201] S.Z. Xu, J.K. Liu, X.M. Li, and Y.S. Ma. A full-scale topology optimization method for surface fiber reinforced additive manufacturing parts. *Computer Methods in Applied Mechanics and Engineering*, 401:115632, 2022.

[202] S.Z. Xu, J.K. Liu, B. Zou, Q.H. Li, and Y.S. Ma. Stress constrained multi-material topology optimization with the ordered SIMP method. *Computer Methods in Applied Mechanics and Engineering*, 373:113453, 2021.

[203] T. Xu, X.D. Huang, X.S. Lin, and Y.M. Xie. Topology optimization for maximizing buckling strength using a linear material model. *Computer Methods in Applied Mechanics and Engineering*, 417:116437, 2023.

[204] M. Yaghmaei, A. Ghoddosian, and M.M. Khatibi. A filter-based level set topology optimization method using a 62-line MATLAB code. *Structural and Multidisciplinary Optimization*, 62:1001–1018, 2020.

[205] T. Yamada, K. Izui, S. Nishiwaki, and A. Takezawa. A topology optimization method based on the level set method incorporating a fictitious interface energy. *Computer Methods in Applied Mechanics and Engineering*, 199(45–48):2876–2891, 2010.

[206] S. Yan, F. Wang, and O. Sigmund. On the non-optimality of tree structures for heat conduction. *International Journal of Heat and Mass Transfer*, 122:660–680, 2018.

[207] X.Y. Yan, Y. Liang, and G.D. Cheng. Discrete variable topology optimization for simplified convective heat transfer via sequential approximate integer programming with trust-region. *International Journal for Numerical Methods in Engineering*, 122(20):5844–5872, 2021.

[208] G.H. Yoon. Structural topology optimization for frequency response problem using model reduction schemes. *Computer Methods in Applied Mechanics and Engineering*, 199(25–28):1744–1763, 2010.

[209] C. Yuhn, Y. Sato, H. Kobayashi, A. Kawamoto, and T. Nomura. 4D topology optimization: Integrated optimization of the structure and self-actuation of soft bodies for dynamic motions. *Computer Methods in Applied Mechanics and Engineering*, 414:116187, 2023.

[210] T. Zeng, H. Wang, M. Yang, and J. Alexandersen. Topology optimization of heat sinks for instantaneous chip cooling using a transient pseudo-3D thermofluid model. *International Journal of Heat and Mass Transfer*, 154:119681, 2020.

[211] X.Y. Zhai, F.L. Chen, and J. Wu. Alternating optimization of design and stress for stress-constrained topology optimization. *Structural and Multidisciplinary Optimization*, 64:2323–2342, 2021.

[212] W.S. Zhang, J. Yuan, J. Zhang, and X. Guo. A new topology optimization approach based on Moving Morphable Components (MMC) and the ersatz material model. *Structural and Multidisciplinary Optimization*, 53:1243–1260, 2016.

[213] W.S. Zhang, W.L. Zhong, and X. Guo. An explicit length scale control approach in SIMP-based topology optimization. *Computer Methods in Applied Mechanics and Engineering*, 282:71–86, 2014.

[214] W.S. Zhang, J.H. Zhou, Y.C. Zhu, and X. Guo. Structural complexity control in topology optimization via moving morphable component (MMC) approach. *Structural and Multidisciplinary Optimization*, 56:535–552, 2017.

[215] X.P. Zhang and Z. Kang. Dynamic topology optimization of piezoelectric structures with active control for reducing transient response. *Computer Methods in Applied Mechanics and Engineering*, 281:200–219, 2014.

[216] J. Zhao, L. Meng, X. Lan, H. Li, L. Gao, and Z. Wang. Design and experimental verification of self-supporting topologies for selective laser melting. *Thin-Walled Structures*, 161:107419, 2021.

[217] L. Zhou and W. Zhang. Topology optimization method with elimination of enclosed voids. *Structural and Multidisciplinary Optimization*, 60:117–136, 2019.

[218] M. Zhou and G.I.N. Rozvany. The COC algorithm, part II: Topological, geometrical and generalized shape optimization. *Computer Methods in Applied Mechanics and Engineering*, 89(1–3):309–336, 1991.

[219] M. Zhou and G.I.N. Rozvany. On the validity of ESO type methods in topology optimization. *Structural and Multidisciplinary Optimization*, 21(1):80–83, 2001.

[220] B. Zhu, X. Zhang, H. Zhang, J. Liang, H. Zang, H. Li, and R. Wang. Design of compliant mechanisms using continuum topology optimization: A review. *Mechanism and Machine Theory*, 143:103622, 2020.

[221] J.H. Zhu, H. Zhou, C. Wang, L. Zhou, S.Q. Yuan, and W.H. Zhang. A review of topology optimization for additive manufacturing: Status and challenges. *Chinese Journal of Aeronautics*, 34(1):91–110, 2021.

[222] C. Zhuang and Z. Xiong. A global heat compliance measure based topology optimization for the transient heat conduction problem. *Numerical Heat Transfer, Part B: Fundamentals*, 65(5):445–471, 2014.

[223] W.J. Zuo and K. Saitou. Multi-material topology optimization using ordered SIMP interpolation. *Structural and Multidisciplinary Optimization*, 55:477–491, 2017.

[224] Z.H. Zuo and Y.M. Xie. A simple and compact Python code for complex 3D topology optimization. *Advances in Engineering Software*, 85:1–11, 2015.

Index

Note: Page numbers in **bold** and *italics* refer to tables and figures, respectively.

For Product Safety Concerns and Information please contact our EU representative GPSR@taylorandfrancis.com Taylor & Francis Verlag GmbH, Kaufingerstraße 24, 80331 München, Germany

Batch number: 10399852

Printed by Printforce, the Netherlands